EXPANDING UNDERREPRESENTED MINORITY PARTICIPATION

Committee on Underrepresented Groups and
the Expansion of the Science and Engineering Workforce Pipeline

Committee on Science, Engineering, and Public Policy
Policy and Global Affairs

NATIONAL ACADEMY OF SCIENCES,
NATIONAL ACADEMY OF ENGINEERING, AND
INSTITUTE OF MEDICINE
OF THE NATIONAL ACADEMIES

THE NATIONAL ACADEMIES PRESS
Washington, D.C.
www.nap.edu

THE NATIONAL ACADEMIES PRESS 500 Fifth Street, N.W. Washington, DC 20001

NOTICE: The project that is the subject of this report was approved by the Governing Board of the National Research Council, whose members are drawn from the councils of the National Academy of Sciences, the National Academy of Engineering, and the Institute of Medicine. The members of the committee responsible for the report were chosen for their special competences and with regard for appropriate balance.

This study was supported by grants between the National Academy of Sciences and the National Science Foundation, National Institutes of Health, National Aeronautics and Space Administration, Carnegie Corporation of New York, and Otto Haas Charitable Trust #2. Any opinions, findings, conclusions, or recommendations expressed in this publication are those of the authors and do not necessarily reflect the views of the organizations or agencies that provided support for the project.

International Standard Book Number-13: 978-0-309-15968-5 (Book)
International Standard Book Number-10: 0-309-15968-7 (Book)
International Standard Book Number-13: 978-0-309-15969-2 (PDF)
International Standard Book Number-10: 0-309-15969-5 (PDF)
Library of Congress Catalog Card Number: 20110=922442

Additional copies of this report are available from the National Academies Press, 500 Fifth Street, N.W., Lockbox 285, Washington, DC 20055; (800) 624-6242 or (202) 334-3313 (in the Washington metropolitan area); Internet, http://www.nap.edu.

THE NATIONAL ACADEMIES
Advisers to the Nation on Science, Engineering, and Medicine

The **National Academy of Sciences** is a private, nonprofit, self-perpetuating society of distinguished scholars engaged in scientific and engineering research, dedicated to the furtherance of science and technology and to their use for the general welfare. Upon the authority of the charter granted to it by the Congress in 1863, the Academy has a mandate that requires it to advise the federal government on scientific and technical matters. Dr. Ralph J. Cicerone is president of the National Academy of Sciences.

The **National Academy of Engineering** was established in 1964, under the charter of the National Academy of Sciences, as a parallel organization of outstanding engineers. It is autonomous in its administration and in the selection of its members, sharing with the National Academy of Sciences the responsibility for advising the federal government. The National Academy of Engineering also sponsors engineering programs aimed at meeting national needs, encourages education and research, and recognizes the superior achievements of engineers. Dr. Charles M. Vest is president of the National Academy of Engineering.

The **Institute of Medicine** was established in 1970 by the National Academy of Sciences to secure the services of eminent members of appropriate professions in the examination of policy matters pertaining to the health of the public. The Institute acts under the responsibility given to the National Academy of Sciences by its congressional charter to be an adviser to the federal government and, upon its own initiative, to identify issues of medical care, research, and education. Dr. Harvey V. Fineberg is president of the Institute of Medicine.

The **National Research Council** was organized by the National Academy of Sciences in 1916 to associate the broad community of science and technology with the Academy's purposes of furthering knowledge and advising the federal government. Functioning in accordance with general policies determined by the Academy, the Council has become the principal operating agency of both the National Academy of Sciences and the National Academy of Engineering in providing services to the government, the public, and the scientific and engineering communities. The Council is administered jointly by both Academies and the Institute of Medicine. Dr. Ralph J. Cicerone and Dr. Charles M. Vest are chair and vice chair, respectively, of the National Research Council.

www.national-academies.org

Staff

Peter H. Henderson, Co-Study Director
Earnestine Psalmonds, Co-Study Director*
Neeraj P. Gorkhaly, Research Associate

*Earnestine Psalmonds is a visiting scholar from the National Science Foundation (NSF). This work was supported by the NSF, but findings, conclusions, or recommendations expressed are those of the authors and do not necessarily reflect the views of the NSF.

Acknowledgment of Reviewers

This report has been reviewed in draft form by individuals chosen for their diverse perspectives and technical expertise, in accordance with procedures approved by the National Academies' Report Review Committee. The purpose of this independent review is to provide candid and critical comments that will assist the institution in making its published report as sound as possible and to ensure that the report meets institutional standards for objectivity, evidence, and responsiveness to the study charge. The review comments and draft manuscript remain confidential to protect the integrity of the process.

We wish to thank the following individuals for their review of this report: Betsy Ancker-Johnson, General Motors Corporation (retired); Carlos Castillo-Chavez, Arizona State University; Daryl Chubin, American Association for the Advancement of Science; Henry Frierson, University of Florida; Adam Gamoran, University of Wisconsin-Madison; Juliet Garcia, University of Texas at Brownsville; Tuajuanda Jordan, Howard Hughes Medical Institute; Shirley Malcom, American Association for the Advancement of Science; Douglas Medin, Northwestern University; Samuel Myers, University of Minnesota; Helen Quinn, SLAC National Accelerator Laboratory; Joan Reede, Harvard University; Isiah Warner, Louisiana State University; and Herman White, Fermi National Accelerator Laboratory.

Although the reviewers listed above have provided many constructive comments and suggestions, they were not asked to endorse the conclusions or recommendations, nor did they see the final draft of the report before its release. The review of this report was overseen by Georges C. Benjamin,

American Public Health Association, and Mary E. Clutter, National Science Foundation (retired). Appointed by the National Academies, they were responsible for making certain that an independent examination of this report was carried out in accordance with institutional procedures and that all review comments were carefully considered. Responsibility for the final content of this report rests entirely with the authoring committee and the institution.

The study committee thanks the National Science Foundation, the National Institutes of Health, the National Aeronautics and Space Administration, the Otto P. Haas Foundation, and the Carnegie Corporation of New York City for the financial support they provided for this study and the many experts who met with the committee to provide their insights on the legal and the policy context, role of diversity in science and engineering and guidance for a national response to broaden participation of under-represented minorities. We also thank the staff of the National Academies who helped organize our committee meetings and draft the report.

Freeman A. Hrabowski, III, *Chair*
Committee on Underrepresented Groups and the Expansion of the Science and Engineering Workforce Pipeline

Contents

Summary   1
Introduction
    1   A Strong Science and Engineering Workforce   17
    2   Dimensions of the Problem   33
Becoming Scientists and Engineers
    3   Preparation   53
    4   Access and Motivation   91
    5   Affordability   103
    6   Academic and Social Support   129
Conclusion
    7   The Journey Beyond the Crossroads   143
    8   Recommendations and Implementation Actions   171
Bibliography   191

APPENDIXES

   A   Charge to the Study Committee   205
   B   U.S. Senate Letter to the National Academy of Sciences   207
   C   Committee Member Biographies   211
   D   Agendas for Public Meetings   225
   E   Recommendations on STEM Education from *Rising Above the Gathering Storm*   233
   F   Ingredients for Success in STEM   239
   G   Baccalaureate Origins of Underrepresented Minority PhDs   249
   H   An Agenda for Future Research   265

TABLES

2-1 Percentage Change in S&E Degrees Earned, by Degree Level and Race/Ethnicity (Bachelor's and Master's 1998-2007; Doctorates 1998-2007), 44

2-2 Principal Investigators on NIH Research Grants, by Race/Ethnicity, 48

2-3 NSF Research Proposals and Awards, by Race/Ethnicity of PI, 2009, 49

3-1 Average Mathematics Scores of Students from Beginning Kindergarten to Grade 8, by Race/Ethnicity: 1998, 2000, 2002, 2004, and 2007, 61

3-2 Average Science Score of Students in Grades 4, 8, and 12, by Race/Ethnicity: 1996, 2000, and 2005, 62

3-3 Average Mathematics Scale Scores and Achievement Level Results by Race/Ethnicity for 4th and 8th Grade Public School Students, 2007, 64

3-4 Average Science Scale Scores by Race/Ethnicity and Grade: 2000 and 2005, 66

3-5 Average Scores on the SAT Reasoning Test by Race/Ethnicity, 2009, 83

3-6 Average State Mathematics Scores on the SAT Reasoning Test by Race/Ethnicity, 2009, 84

5-1 Primary Support Mechanism for S&E Doctorate Recipients, by Citizenship, Sex, and Race/Ethnicity: 2007, 116

5-2 Primary Mechanisms of Support for S&E Doctorate Recipients by Citizenship, Sex, and Race/Ethnicity: 2005, 124

7-1 Approaches to Increasing Underrepresented Minority Participation and Success in Science and Engineering, 145

7-2 Number of Baccalaureate Institutions of African American PhDs in Science and Engineering, by Broad Field and Institutional Type, 2006, 151

8-1 Cost Estimates for New Underrepresented Minority Student Support (millions of dollars), 188

G-1 Top 25 Baccalaureate Origin Institutions of African American Doctorates in the Natural Sciences and Engineering (NS&E) 2002-2006, 252

G-2   Top 15 Baccalaureate Origin Institutions of African American Doctorates in the Natural Sciences and Engineering (NS&E) That Are Historically Black Colleges and Universities (HBCUs), by Broad Field, 2002-2006 (most recent five years), 254

G-3   Top 13 Baccalaureate Origin Institutions of African American Doctorates in the Natural Sciences and Engineering (NS&E) That Are Predominantly White Universities, by Broad Field, 2002-2006 (most recent five years), 256

G-4   Top 25 Baccalaureate Origin Institutions of Hispanic Doctorates in the Natural Sciences and Engineering (NS&E), 2002-2006, 261

G-5   Top 25 Baccalaureate Origin Institutions of Hispanic Doctorates in the Natural Sciences and Engineering (NS&E), by Broad Field, 2002-2006 (most recent five years), 260

FIGURES

1-1   U.S. Population and U.S. Science and Engineering Workforce, by Race/Ethnicity, 2006, 25

1-2   U.S. Population by Race/Ethnicity, 1990-2050 (2010-2050 projected), 26

2-1   Enrollment and Degrees, by Educational Level and Race/Ethnicity/Citizenship, 2007, 37

2-2   Trends in Students' Aspiration to Major in a STEM Discipline by Racial Identification, 1971-2009, 40

2-3   Percentage of 2004 STEM Aspirants Who Completed STEM Degrees in Four and Five Years, by Race/Ethnicity, 41

2-4   Four- and Five-Year Degree Completion Rates of 2004 Freshmen, by Initial Major Aspiration and Race/Ethnicity, 42

2-5   Underrepresented Minorities Among S&E Degree Recipients, by Degree Level, 2006, 45

2-6   Temporary Residents Among S&E Degree Recipients, by Degree Level, 2006, 46

2-7   Doctoral Scientists and Engineers Employed in Four-Year Institutions, by Race/Ethnicity, 2006, 48

3-1   Percentage Distribution of Public School Students Enrolled in Kindergarten Through 12th Grade by Race/Ethnicity: Selected Years, October 1972-October 2007, 57

3-2   Grade 8 TIMSS Average Math Scores by Race/Ethnicity, 58

3-3   Grade 8 TIMSS Average Math Scores by School Poverty Level, 58

3-4   TIMSS Grade 4 Math Racial/Ethnic Subgroup Comparison to All Participating Countries, 59

3-5 Percentage of Students with Core Course Work During High School by Race/Ethnicity, 1999 and 2009, 85

3-6 Access to AP by Race/Ethnicity—U.S. Public Schools: High School Class of 2009, 86

3-7 Percentage of High School Students Taking Pre-Calculus by Race/Ethnicity: 1999 and 2009, 87

4-1 Fall Undergraduate Enrollment in Postsecondary Institutions, by Race/Ethnicity, 1976-2004, 92

5-1 Source of Financial Aid Received by Undergraduates: 2007-2008, 107

5-2 Full-Time S&E Graduate Students by Field and Mechanism of Support: 2006, 115

5-3 U.S. Citizen and Permanent Resident Doctorate Recipients with Levels of Graduate School Debt Greater Than $30,000, by Broad Field of Study and Race/Ethnicity, 2008, 121

5-4 Graduate Coursework, Degrees Pursued, and Degrees Completed, LSAMP Participants Compared to National Underrepresented Minorities and National White and Asian American Graduates, 123

BOXES

1-1 Grand Challenges for Engineering, 18

1-2 Science, The Endless Frontier, 19

1-3 Lost at the Frontier: U.S. Science and Technology Policy Adrift, 20

1-4 The Context for Innovation and Competitiveness Policy, 21

3-1 Education Goals 2000, 55

3-2 Knowledge Is Power (KIPP) Program, 67

3-3 For Inspiration and Recognition of Science and Technology (FIRST) Program, 68

3-4 Advancement Via Individual Determination (AVID) Program, 69

3-5 Indigenous Education Institute, 70

3-6 The El Paso Collaborative for Academic Excellence, 71

4-1 No Longer Separate, Not Yet Equal, 96

4-2 Aiming High, 96

4-3 Why African American Students Should Major in Biomedical Research, 101

5-1   Financial Support of Doctoral Completion, 118

6-1   A Seven-Step Plan to Lower College Dropout Rates, 131
6-2   Broadening Participation in Graduate School, 132

7-1   Rice University Computational and Applied Mathematics
        Program, 152
7-2   Minority-Serving Institutions, 155
7-3   Life-Gets-Better at Florida A&M University, 157
7-4   Windows of Opportunity, Miami Dade College, 161
7-5   Selected Promising Interventions, 166
7-6   Review of Literature on Student Support, 168

F-1   Four Key Strands in K-8 Science Education, 242
F-2   Vision and Change in Undergraduate Biology Education: A Call to
        Action, 243

Summary

A TRANSFORMATIONAL MOMENT

Our ability to meet the challenges and achieve the opportunities of our time depends in large measure on our science and engineering (S&E) enterprise. Yet, while our S&E capability is as strong as ever, the dominance of the United States in these fields has lessened as the rest of the world has invested in and grown their research and education capacities. *Rising Above the Gathering Storm* documented this global leveling and argued that the United States was at a **crossroads**: For the United States to maintain the global leadership and competitiveness in science and technology that are critical to achieving national goals today, we must invest in research, encourage innovation, and grow a strong, talented, and innovative science and technology workforce.[1] *Gathering Storm* resonated strongly in both the executive and legislative branches of government, resulting in the American Competitive Incentive Act, the America COMPETES Act, and substantial appropriations through the American Recovery and Reinvestment Act of 2009.

The importance of S&E to the United States has been documented in a series of reports over more than half a century. Nevertheless, critical issues for the nation's S&E infrastructure remain unsettled. Among them, America faces a demographic challenge with regard to its S&E workforce: Minorities are seriously underrepresented in science and engineering, yet

---

[1] Institute of Medicine, National Academy of Sciences, and National Academy of Engineering. 2007. *Rising Above the Gathering Storm: Energizing and Employing America for a Brighter Economic Future.* Washington, DC: The National Academies Press.

1

they are also the most rapidly growing segment of the population. *Gathering Storm* provided compelling recommendations for sustaining and increasing our knowledge workforce as part of a larger plan to sustain the nation's scientific and technological leadership. These workforce recommendations focused on improving K-12 STEM education as well as providing incentives for students to pursue S&E education at the undergraduate and graduate levels.[2] We fully support these recommendations, but they are insufficient to meet the emerging demographic realities. The United States stands again at the **crossroads**: A national effort to sustain and strengthen S&E must also include a strategy for ensuring that we draw on the minds and talents of all Americans, including minorities who are underrepresented in S&E and currently embody a vastly underused resource and a lost opportunity for meeting our nation's technology needs.

Citing the need to develop a strong and diverse S&E workforce, U.S. Senators Edward Kennedy, Barbara Mikulski, Patty Murray, and Hillary Clinton requested in November 2006 a study of underrepresented minority participation in S&E. The U.S. Congress later included this request as a mandate in the 2007 America COMPETES Act, charging the study committee to explore the role of diversity in the STEM workforce and its value in keeping America innovative and competitive, analyze the rate of change and the challenges the nation currently faces in developing a strong and diverse workforce, and identify best practices and the characteristics of these practices that make them effective and sustainable.

AMERICA'S SCIENCE AND ENGINEERING TALENT AT THE CROSSROADS

Broad Participation Matters

A strategy to increase the participation of underrepresented minorities in science and engineering should play a central role in our approach to sustaining America's research and innovation capacity for at least three reasons:

1. Our sources for the future S&E workforce are uncertain: For many years, the nation relied on an S&E workforce that was predominantly male and overwhelmingly white and Asian. In the more recent past, as the proportion of the S&E workforce that is white and male has fluctuated, we have seen gains for women in some fields and an increasing reliance on international students in others. Non-U.S. citizens, particularly those from China and India, have accounted for almost all growth in STEM doctorate

---

[2] Ibid. pp. 5-7, 9-10.

awards and in some engineering fields comprise the majority of new doctorates. Yet, we are coming to understand that relying on non-U.S. citizens for our S&E workforce is an increasingly uncertain proposition.

2. *The demographics of our domestic population are shifting dramatically:* If the uncertainty about the future participation of international students suggests that we need to ensure that we draw on all demographic sources, the dramatic changes in the demographics of the domestic population, especially the school-age population, suggest that the problem is all the more urgent: *Those groups that are most underrepresented in S&E are also the fastest growing in the general population.*

3. *Diversity is an asset:* Increasing the participation and success of underrepresented minorities in S&E contributes to the health of the nation by expanding the S&E talent pool, enhancing innovation, and improving the nation's global economic leadership.

Dimensions of the Problem

The S&E workforce is large and fast-growing: more than 5 million strong and projected by the U.S. Bureau of Labor Statistics to grow faster than any other sector in coming years. This growth rate provides an opportunity as well as an obligation to draw on new sources of talent to make the S&E workforce as robust and dynamic as possible. But we start from a challenging position: Underrepresented minority groups comprised 28.5 percent of our national population in 2006, yet just 9.1 percent of college-educated Americans in science and engineering occupations (academic and nonacademic), suggesting the proportion of underrepresented minorities in S&E would need to *triple* to match their share of the overall U.S. population.

Underrepresentation of this magnitude in the S&E workforce stems from the underproduction of minorities in S&E at every level of postsecondary education, with a progressive loss of representation as we proceed up the academic ladder. In 2007, underrepresented minorities comprised 38.8 percent of K-12 public enrollment, 33.2 percent of the U.S college age population, 26.2 percent of undergraduate enrollment, and 17.7 percent of those earning science and engineering bachelor's degrees. In graduate school, underrepresented minorities comprise 17.7 percent of overall enrollment but are awarded just 14.6 percent of S&E master's degrees and a miniscule 5.4 percent of S&E doctorates.

Historically, there has been a strong connection between increasing educational attainment in the United States and the growth in and global leadership of the economy. Consequently, there have been calls—from the College Board, the Lumina and Gates Foundations, and the administration—to increase the postsecondary completion rate in the United

States from 39 percent to 55 or 60 percent. The challenge is greatest for underrepresented minorities: In 2006 only 26 percent of African Americans, 18 percent of American Indians, and 16 percent of Hispanics in the 25- to 29-year-old cohort had attained at least an associate degree.[3] The news is even worse in S&E fields. In 2000, as noted in *Gathering Storm,* the United States ranked 20 out of 24 countries in the percentage of 24-year-olds who had earned a first degree in the natural sciences or engineering. Based on these data, *Gathering Storm* recommended efforts to increase the percentage of 24-year-olds with these degrees from 6 percent to at least 10 percent, the benchmark already attained by several countries.[4] But again, the statistics are even more alarming for underrepresented minorities. These students would need to *triple, quadruple,* or even *quintuple* their proportions with a first university degree in these fields in order to achieve this 10 percent goal: At present, just 2.7 percent of African Americans, 3.3 percent of Native Americans and Alaska Natives, and 2.2 percent of Hispanics and Latinos who are 24 years old have earned a first university degree in the natural sciences or engineering.[5]

Recent data from the Higher Education Research Institute (HERI) at UCLA show that underrepresented minorities aspire to major in STEM in college at the same rates as their white and Asian American peers, and have done so since the late 1980s. Yet, these underrepresented minorities have lower four- and five-year completion rates relative to those of whites and Asian Americans. That a similar picture previously was seen in data in the mid-1990s signals that, although we have been aware of these problems for some time, we, as a nation, have made little collective progress in addressing them.

Fixing the Problem

No single career pathway or pipeline exists in STEM education. Students start from diverse places, with different family backgrounds and schools and communities with different resources and traditions. There also is substantial variation in K-12 mathematics and science education across schools, districts, and states. STEM courses, moreover, serve varied purposes for students on different tracks.

---

[3] Ryu Mikyung. 2008. *Minorities in Higher Education.* Washington, DC: American Council on Education.

[4] IOM, NAS, and NAE. 2007. *Rising Above the Gathering Storm: Energizing and Employing America for a Brighter Economic Future.* Washington, DC: The National Academies Press.

[5] National Science Foundation, Women, Minorities, and Persons with Disabilities in Science and Engineering, http://www.nsf.gov/statistics/wmpd/ (accessed March 27, 2009); and U.S. Census Bureau, Population estimates, http://www.census.gov/popest/national/asrh/NC-EST2007-asrh.html (accessed March 27, 2009).

Although a set of pathways may be difficult to describe in detail, the ingredients for success in STEM are the acquisition of knowledge, skills, and habits of mind; opportunities to put these into practice; a developing sense of competence and progress; motivation to be in, a sense of belonging to, or self-identification with the field; and information about stages, requirements, and opportunities. These ingredients require attention in some measure for all students at every stage along the STEM educational continuum. However, there are issues that are specific to underrepresented minorities, in general and in STEM, focused on preparation, access and motivation, financial aid, academic support, and social integration.

Preparation

The education children receive from preschool through high school is foundational and critical. For STEM, quality preparation is a prerequisite for later success. From "A Nation at Risk" 25 years ago to current debates over reauthorization of the No Child Left Behind Act, interventions have been a subject of contention. Yet today, the nation remains faced with many of the same issues it has grappled with for years: failing schools, inequitable distributions of resources across schools, achievement gaps, and increasing demand for skilled workers in science, technology, and other knowledge-intensive fields. Moreover, substantial growth in the nation's Hispanic population has increased pressure on our nation's schools by increasing the number of nonnative English speakers.

Researchers offer many explanations for the persistent achievement gaps while recognizing that there are many interrelated factors. They agree that family and community differences, school context, low expectations, and lack of exposure to role models, information about career opportunities, and advanced courses affect minority students' success in mathematics and science. Although there is considerable disagreement over solutions such as school choice, testing, and teacher pay, there is substantial agreement about the need for strong preschool programs, more qualified mathematics and science teachers in predominantly minority and low-income schools, and challenging high school curricula that prepare underrepresented minorities for college.

Access and Motivation

The S&E workforce in the United States is drawn primarily from among our nation's undergraduates who complete at least a bachelor's degree. Undergraduate enrollment of underrepresented minorities has increased substantially over the past three decades and at a rate faster than for whites. As a result, they now comprise 26.2 percent of all undergraduates. While this falls short of their proportion in the college age population (33.2 percent), this increase in their numbers and proportions nonetheless represents a significant national achievement.

However, we must do much more to attract and retain underrepresented minorities, low-income students, and first-generation undergraduates who aspire to a major in STEM. Specifically, we can do the following: (1) improve college awareness activities for prospective college students, (2) focus on college admissions policies that support the postsecondary matriculation of qualified underrepresented minority students, (3) raise awareness of STEM careers through K-12 activities, improved counseling for science and mathematics, and activities that promote STEM, and (4) promote STEM outreach that specifically targets underrepresented minorities.

Affordability

College affordability is an issue for all students, especially as tuition continues to increase above the rate of inflation, and is affected by federal, state, and institutional policies. Financial support that meets student need is strongly correlated with student attendance and persistence. For underrepresented minorities in STEM, financial support can come from a range of programs, including need-based financial aid programs (e.g., Pell Grants), general programs supporting underrepresented minorities (e.g., Gates Millennium Scholarships), financial aid that targets students in STEM (e.g., SMART Grants), and programs that target underrepresented minorities in STEM (e.g., NIH's MARC program). While some financial assistance may be need-based, programs that target underrepresented minorities in STEM are necessary. Researchers have shown that financial incentives are most effective in reducing attrition among low-income and minority students when provided in conjunction with academic support and campus integration programs.

Academic and Social Support

A study of undergraduate persistence by the National Center for Education Statistics (NCES) found that although women were less likely to major in STEM than men, they had similar or higher persistence rates. By contrast, they found that underrepresented minorities majored in STEM at the same rate as others, but their completion rate was lower, a finding recently corroborated by HERI. NCES concluded that underrepresented minorities faced greater barriers to persistence and completion. Other researchers note also that the culture and climate of institutions, including the diversity of faculty, impact the entire process from entry to graduation.

Several practical steps can be taken to increase the completion of minorities: Make student success a priority, track student achievement, identify "choke points" such as course availability, make course transfer easier, and ensure that courses are structured to properly support students. Only higher education institutions can address these issues and only they can ensure the academic and social support necessary for underrepresented minority students in STEM. To address issues of self-confidence and inclusion that are profoundly salient, institutions can play a pivotal role, through formal and informal actions, to encourage persistence through:

- Strong leadership from trustees and regents, the president, provost, deans, and department chairs;
- A campus-wide commitment to inclusiveness;
- A deliberate process of self-appraisal focused on campus climate;
- Development of a plan to implement constructive change; and
- Ongoing evaluation of implementation efforts.

THE JOURNEY BEYOND THE CROSSROADS

Principles

Six principles have informed the development of our recommendations to move "beyond the crossroads" to the implementation of actions designed to increase the participation and success of underrepresented minorities in STEM education. Given how long it takes to realize gains from educational reform, the national effort must be urgent, sustained, comprehensive, intensive, coordinated, and informed:

1. The problem is *urgent* and will continue to be for the foreseeable future.
2. A successful national effort to address underrepresented minority participation and success in STEM will be *sustained*.
3. The potential for losing students along all segments of the pathway from preschool through graduate school necessitates a *comprehensive* approach that focuses on all segments of the pathway, all stakeholders, and the potential of all programs, targeted or nontargeted.
4. Students who have not had the same degree of exposure to STEM and to postsecondary education require more *intensive* efforts at each level to provide adequate preparation, financial support, mentoring, social integration, and professional development.
5. A *coordinated* approach to existing federal STEM programs can leverage resources while supporting programs tailored to the specific mis-

sions, histories, cultures, student populations, and geographic locations of institutions with demonstrated success.

6.   Evaluation of STEM programs and increased research on the many dimensions of underrepresented minorities' experience in STEM help ensure that programs are well *informed,* well designed, and successful.

Institutional Roles

The diversity of American higher education institutions is a competitive advantage in the global knowledge economy as different types of institutions address the varied needs of students who find themselves at different places in their educational journey with a range of life and career goals. This institutional diversity could be, but is not yet, effective in addressing the varied needs of the nation's underrepresented minority students in STEM. For our recommended action to be successful, every institution of higher education should take steps to address the problem of underrepresented minority participation in STEM. Currently, only a small number of institutions are doing so. They are diverse and can be found among all institutional types and categories; they are successful because they are doing something special to support the retention and completion of underrepresented minority undergraduates in the natural sciences and engineering. Their actions can be replicated, and when they are, with a focus on both numbers and quality, it will pay off significantly:

•   **PredominantlyWhite Institutions:** The best way to increase the **retention** of underrepresented minorities in STEM is to replicate programs of the successful PWIs at *a very large number* of similar institutions, especially large state flagships.

•   **Minority-Serving Institutions:** MSIs have a legacy of recruiting, retaining, and graduating a disproportionate number of minorities, especially at the undergraduate level. With additional support, MSIs can expand their effectiveness in recruiting, retaining, and graduating an increased number of minorities, especially at the baccalaureate level.

•   **Community Colleges:** To facilitate and increase the successful transfer of underrepresented minorities in STEM to four-year institutions, an increased emphasis on and support for articulation agreements, summer bridge programs, mentoring, academic and career counseling, peer support, and undergraduate research at two-year institutions are recommended.

Leadership

Leadership is key to the successful transformation of institutions and the development of sustainable programs:

- **Sectoral Leadership:** Leadership in identifying and articulating minority participation and success as an institutional goal is essential at all levels for all stakeholders: the federal government, state and local governments, employers, philanthropy, professional societies, educational institutions, programs, faculty, and students.
- **Institutional Leadership:** At each higher education institution, the academic leadership—regents, trustees, presidents, provosts, deans, and department chairs—should articulate underrepresented minority participation as a key commitment to set a tone that raises awareness and effort. Faculty buy-in is essential. Institutional leaders also should be more aggressive in investing in the development of underrepresented minority teachers, faculty, and administrators who can serve as role models and leaders.
- **Programmatic Leadership:** A champion at the program level providing leadership dedicated to long-term improvement is typically critical to the success of underrepresented minority programs at the undergraduate and graduate levels.

Program Development

The literature on best practices for increasing minority participation in STEM education provides guidance for the development and execution of the policies and programs that are designed to change the academic culture and sustain programs so as to encourage student retention, persistence, and completion. Below are key elements for developing a program that are necessary to transform goals into reality.

- **Resources and Sustainability:** The development of programs to stimulate student interest and success in STEM, in general and for programs that target minorities, requires substantial and sustained resources.
- **Coordination and Integration:** Coordination and integration of efforts can make the aggregate of individual programs greater than the sum of their parts.
- **Focus on the Pipeline, Career Pathways, and Transition Points:** A corollary to coordination and integration is programs and strategies that focus on career pathways and critical pipeline transition points.
- **Program Design:** A successful program may be innovative or replicative and will draw on the lessons of best and worst practices in program development and implementation, but it will be tailored to its particular institutional and disciplinary context.
- **Program Execution:** Even if a program is well designed, well resourced, and appropriately targeted, without proper execution it has little chance of full success.

- **Program Evaluation:** Whether a program meets or exceeds organizational goals is subject to examination. Programs designed to increase the participation of underrepresented minorities benefit themselves and others by engaging in ongoing, constructive evaluation.
- **Knowledge Sharing:** A corollary to the importance of program evaluation is the dissemination of information about practice derived from these evaluations and other research.

Program Characteristics

While many strategies for academic support and social integration apply equally to students in STEM fields regardless of their racial or ethnic background, for underrepresented minority students these can be critical for opening doors of opportunity. Proven, intensive interventions for underrepresented minorities in STEM include:

- **Summer Programs:** Summer programs that include or target minority middle and high school and undergraduate students provide experiences that stimulate interest in these fields through study, hands-on research, and the development of a cadre of students who support each other in their interests.
- **Research Experiences:** At the undergraduate and graduate level, engagement in rich research experiences allows for the further development of interest and competence in and identification with STEM and enhances academic competitiveness.
- **Professional Development Activities:** Opportunities for undergraduate and graduate students to engage in networking, participation in conferences, and presentation of research provide opportunities to develop and socialize students within a discipline and profession.
- **Academic Support and Social Integration:** Success may also hinge on the extent to which undergraduate and graduate students participate in activities—such as peer-to-peer support, study groups, social activities, tutoring, and mentoring programs—that can promote academic success and social integration.
- **Mentoring:** Engaged mentors can provide undergraduate and graduate students with information, advice, and guidance and support generally and at critical decision points.

Students should also have access to proper facilities and equipment, and course curricula should be formulated to encourage student learning and progress—something that seems self-evident, except that many introductory courses in the sciences have traditionally sought to "weed out" students rather than encourage them.

Recommendations and Implementation Actions

A successful national effort to increase the participation and success of underrepresented minorities in STEM will be urgent, sustained, comprehensive, intensive, coordinated, and informed. It will also cut across all educational stages and stakeholder groups. With these principles in mind, the committee has developed six broad recommendations followed by implementation actions that should be taken by specific stakeholders. Following the six broad recommendations, we propose two top priorities that should serve as the near-term focal point for national policies for broadening participation.

Preparation

Recommendation 1: *Preschool through Grade 3 Education*
Prepare America's children for school through preschool and early education programs that develop reading readiness, provide early mathematics skills, and introduce concepts of creativity and discovery.

Recommendation 2: *K to 12 Mathematics and Science*
Increase America's talent pool by vastly improving K-12 mathematics and science education for underrepresented minorities.

Recommendation 3: *K-12 Teacher Preparation and Retention*
Improve K-12 mathematics and science education for underrepresented minorities overall by improving the preparedness of those who teach them those subjects.

Postsecondary Success

Recommendation 4: *Access and Motivation*
Improve access to all postsecondary education and technical training and increase underrepresented minority student awareness of and motivation for STEM education and careers through improved information, counseling, and outreach.

Recommendation 5: *Affordability*
Develop America's advanced STEM workforce by providing adequate financial support to underrepresented minority students in undergraduate and graduate STEM education.

Recommendation 6: *Academic and Social Support*
Take coordinated action to transform the nation's higher education institutions to increase inclusion of and college completion and success in STEM education for underrepresented minorities.

Top-Priority Actions

Out of the recommendations and implementation actions that span the entire educational system and full spectrum of stakeholders, we have identified two areas of highest priority for near-term action. We chose them because we believe they can have the most immediate impact on the critical transition points in the STEM education pathway for underrepresented minorities.

Priority 1: Undergraduate Retention and Completion: We propose, as a short-term focus for increasing the participation and success of underrepresented minorities in STEM, policies and programs that seek to increase undergraduate retention and completion through strong academic, social, and financial support. Financial support for underrepresented minorities that allows them to focus on and succeed in STEM will increase completion and better prepare them for the path ahead. This financial assistance should be provided through higher education institutions along with programs that simultaneously integrate academic, social, and professional development.

The success of such an effort is made possible by the existence of a cadre of qualified underrepresented minorities who already attend college, declared an interest in majoring in the natural sciences or engineering, and either did not complete a degree or switched out of STEM before graduating. An increase in the STEM completion rate for these students may, by example, increase interest in STEM on the part of younger cohorts and also increase the number of underrepresented minorities who may consider graduate education in STEM.

Financial support for underrepresented minorities that allows them to focus on and succeed in STEM will increase completion and better prepare them for the path ahead. This financial assistance should be provided through higher education institutions along with programs that simultaneously provide academic support, social integration, and professional development. Given the scale of the problem, an effort to double the number of underrepresented minorities who complete undergraduate STEM degrees is a near-term, reasonable, and attainable down payment on a longer-term effort to achieve greater parity overall.

Priority 2: Teacher Preparation, College Preparatory Programs, and Transition to Graduate Study: We propose also an emphasis on teacher prepara-

tion, secondary school programs that support preparation for college STEM education, and programs that support the transition from undergraduate to graduate work.

We note the particular importance at the K-12 level of teacher preparation and secondary school programs that support preparation for college STEM education. Secondary school programs that ensure students have access to advanced courses and proper academic advising will support the goal of undergraduate persistence and completion by ensuring that matriculating freshmen are fully prepared for college study.

At the other end of the undergraduate years, programs that support the transition from undergraduate to graduate work are likewise important. The transition of underrepresented minorities to graduate work at top research universities where they can contribute to research and leadership in our nation's science and engineering enterprise is also critical. A significant proportion of new graduate students who are supported through portable fellowships, research assistantships, or institutional grants should be underrepresented minorities in order to increase their overall representation and to move greater numbers into top graduate programs.

INTRODUCTION

1

A Strong Science and Engineering Workforce

ORIGINS OF THE STUDY

Many developing trends of the twenty-first century raise concerns about whether the U.S. science and engineering (S&E) enterprise—the collection of science- and technology-based industries and organizations, federal agencies, and educational institutions—can respond effectively to the challenges and opportunities. We are confronted by pandemics, terrorism, and natural disasters. We are challenged by the need for reliable and affordable energy and a cleaner global environment. We seek a healthier America with greater access to care, more effective medicines, and support for an aging population. We demand strong security, at home and abroad. We aim to develop new products and services for our consumers and to compete in the global marketplace. (See Box 1-1, Grand Challenges for Engineering.)

The importance of S&E to the United States has been documented in a series of reports over more than half a century, from Vannevar Bush's *Science, The Endless Frontier* (1945) to Deborah Shapley and Rustum Roy's *Lost at the Frontier* (1985) to the National Academies' *Rising Above the Gathering Storm* (2007). (See Boxes 1-2 and 1-3.) Yet, while our capability in science and engineering is as strong as ever, the dominance of the United States in these fields has faded as the rest of the world has invested and grown in research and education capacities. *Gathering Storm* documented this global leveling and argued that the United States is at a crossroads: For the United States to maintain the global leadership and competitiveness in science and technology that are critical to achieving national goals today, we

17

BOX 1-1
Grand Challenges for Engineering

In the century just ended, engineering recorded its grandest accomplishments. The widespread development and distribution of electricity and clean water, automobiles and airplanes, radio and television, spacecraft and lasers, antibiotics and medical imaging, and computers and the Internet are just some of the highlights from a century in which engineering revolutionized and improved virtually every aspect of human life. Find out more about the great engineering achievements of the 20th century from a separate NAE Web site: *www.greatachievements.org*.

For all of these advances, though, the century ahead poses challenges as formidable as any from millennia past. As the population grows and its needs and desires expand, the problem of sustaining civilization's continuing advancement, while still improving the quality of life, looms more immediate. Old and new threats to personal and public health demand more effective and more readily available treatments. Vulnerabilities to pandemic diseases, terrorist violence, and natural disasters require serious searches for new methods of protection and prevention. And products and processes that enhance the joy of living remain a top priority of engineering innovation, as they have been since the taming of fire and the invention of the wheel.

In each of these broad realms of human concern—sustainability, health, vulnerability, and joy of living—specific grand challenges await engineering solutions. The world's cadre of engineers will seek ways to put knowledge into practice to meet these grand challenges. Applying the rules of reason, the findings of science, the aesthetics of art, and the spark of creative imagination, engineers will continue the tradition of forging a better future.

—Introduction to *The Grand Challenges for Engineering*, Grand Challenges for Engineering Web site, National Academy of Engineering (2008).

---

must invest in research, encourage innovation, and grow a strong, talented, and innovative science and technology workforce.[1]

This call to action in *Gathering Storm* resonated strongly in both national political parties and in the executive and legislative branches of government, resulting in the American Competitiveness Initiative, the Academic Competitiveness Council, the America COMPETES Act, and spending provisions of the American Recovery and Reinvestment Act. In passing the America COMPETES Act in the summer of 2007, Congress laid the groundwork for the implementation of many of the recommendations from *Gathering Storm*.[2] In passing the American Recovery and Reinvestment Act

---

[1] IOM, NAS, and NAE. 2007. *Rising Above the Gathering Storm: Energizing and Employing America for a Brighter Economic Future*. Washington, DC: The National Academies Press.

[2] America Creating Opportunities to Meaningfully Promote Excellence in Technology, Education, and Science Act, P. L. No. 110-69.

BOX 1-2
Science: The Endless Frontier

One of our hopes is that after the war there will be full employment. To reach that goal the full creative and productive energies of the American people must be released. To create more jobs we must make new and better and cheaper products. We want plenty of new, vigorous enterprises. But new products and processes are not born full-grown. They are founded on new principles and new conceptions which in turn result from basic scientific research. Basic scientific research is scientific capital. Moreover, we cannot any longer depend upon Europe as a major source of this scientific capital. Clearly, more and better scientific research is one essential to the achievement of our goal of full employment.

How do we increase this scientific capital? First, we must have plenty of men and women trained in science, for upon them depends both the creation of new knowledge and its application to practical purposes. We shall have rapid or slow advance on any scientific frontier depending on the number of highly qualified and trained scientists exploring it. . . .

The government should accept new responsibilities for promoting the flow of new scientific knowledge and the development of scientific talent in our youth. These responsibilities are the proper concern of the government, for they vitally affect our health, our jobs, and our national security. It is in keeping also with basic United States policy that the government should foster the opening of new frontiers and this is the modern way to do it.

—From Vannevar Bush, *Science: The Endless Frontier*, a report to the President, July 1945.

---

of 2009 (the Stimulus Act), Congress provided the funding necessary to move forward with the recommendations. (The excerpt from *Rising Above the Gathering Storm* in Box 1-4 provides a description of the innovation and competitiveness policy context. For the education and workforce recommendations of *Rising Above the Gathering Storm*, see Appendix E.)

These topics are not new. In *Educating Americans for the 21st Century* (1983) the National Science Board Commission on Precollege Education in Mathematics, Science and Technology presented a plan of action for improving mathematics, science, and technology education for all American elementary and secondary students and articulated the need for well-trained, highly qualified teachers of mathematics in a technological society.

Nevertheless, critical issues for the nation's S&E infrastructure remain unsettled, in particular the future strength of our nation's science and engineering workforce in light of demographic trends in both the U.S. population and the science and engineering workforce. The *Gathering Storm* provided compelling recommendations for sustaining and increasing our knowledge

---

BOX 1-3
Lost at the Frontier: U.S. Science and Technology Policy Adrift

A standard defense of U.S. academic science is that the university science system gives excellent training to graduate students and postdocs embarking on their careers. But an increasing number of young U.S. scientists are deciding *not* to go to graduate school in the "hard" (or physical) sciences. There has been a decline in the number of bachelor of engineering students who go on to graduate school. The number of M.D.s who go on to get their PhDs has been declining too. So while some leaders brag about our fine university system, young Americans are voting otherwise with their feet.

The trends are different for different fields. Nonetheless, the curves go downward, even in the fields where total graduate enrollments are increasing as a result of the influx of foreign graduate students. There is some debate about the foreign students and their impact on the campus and the scientific workforce, but less attention is being paid to the alarming decline of U.S. citizens seeking advanced training in the physical sciences. . . . Clearly, if bright young Americans continue to be "turned off" university research, the consequences will be serious for the nation.

—From D. Shapley and R. Roy. 1985. *Lost at the Frontier: U.S. Science and Technology Policy Adrift.* Philadelphia, PA: ISI Press.

---

workforce as part of a larger plan to sustain our scientific and technological leadership. These workforce recommendations focused on improving K-12 STEM education as well as providing incentives to students to pursue S&E education at the undergraduate and graduate levels.[3] However, the recommendations are insufficient: A national effort to sustain and strengthen our science and engineering workforce must also include a strategy for ensuring that we draw on the minds and talents of all Americans, including minorities who are underrepresented in science and engineering and currently embody an underused resource and a lost opportunity.

BROAD PARTICIPATION MATTERS

The nation has an opportunity to address simultaneously both our need for a strengthened STEM workforce and the need to respond to the underrepresentation of racial and ethnic minorities in that workforce. This report therefore describes demographic trends in the U.S. population and STEM education that lie metaphorically not only at the S&E crossroads but at the intersection of two quintessentially American stories:

---

[3] NAS, NAE, and IOM. 2007. *Rising Above the Gathering Storm*, pp. 5-7, 9-10.

BOX 1-4
The Context for Innovation and Competitiveness Policy

The United States takes deserved pride in the vitality of its economy, which forms the foundations of our high quality of life, our national security, and our hope that our children and grandchildren will inherit ever greater opportunities. That vitality is derived in large part from the productivity of well-trained people and the steady stream of scientific and technical innovations they produce. Without high-quality, knowledge-intensive jobs and the innovative enterprises that lead to discovery and new technology, our economy will suffer and our people will face a lower standard of living. Economic studies conducted even before the information-technology revolution have shown that as much as 85 percent of measured growth in U.S. income per capita was due to technological change.

Today, Americans are feeling the gradual and subtle effects of globalization that challenge the economic and strategic leadership that the United States has enjoyed since World War II. A substantial portion of our workforce finds itself in direct competition for jobs with lower-wage workers around the globe, and leading-edge scientific and engineering work is being accomplished in many parts of the world. Thanks to globalization, driven by modern communications and other advances, workers in virtually every sector must now face competitors who live just a mouse-click away in Ireland, Finland, China, India, or dozens of other nations whose economies are growing. This has been aptly referred to as "the Death of Distance."

Having reviewed trends in the United States and abroad, the committee is deeply concerned that the scientific and technological building blocks critical to our economic leadership are eroding at a time when many other nations are gathering strength. Although the U.S. economy is doing well today, current trends indicate that the United States may not fare as well in the future without government intervention. This nation must prepare with great urgency to preserve its strategic and economic security. Because other nations have, and probably will continue to have, the competitive advantage of a low wage structure, the United States must compete by optimizing its knowledge-based resources, particularly in science and technology, and by sustaining the most fertile environment for new and revitalized industries and the well-paying jobs they bring.

—From the National Academy of Sciences, National Academy of Engineering, and National Institute of Medicine. 2007. *Rising Above the Gathering Storm: Energizing and Employing America for a Brighter Economic Future.* pp. 1-4.

• The evolution of education in the United States and its role in preparing a workforce that can drive technological innovation and our ability to meet national goals, and

• The stories of African Americans, Hispanics and Latinos, and our nation's native peoples—Native Americans, Alaska Natives, Native

Hawaiians, and Pacific Islanders[4]—who are a growing share of the U.S. population.

A strategy to increase the participation of underrepresented minorities in science and engineering must play a central role in our overall approach to sustaining our capacity to conduct research and innovate. At least three reasons underscore the need for doing so: Our sources for the future S&E workforce are uncertain; the demographics of our domestic population are shifting dramatically; and diversity in S&E is a strength that benefits both diverse groups and the nation as a whole.

Sources of Talent

For many years, the nation has relied on an S&E workforce that has been predominantly male and overwhelmingly white and Asian. In the more recent past, as the proportion of our S&E workforce that is white and male has fluctuated, we have seen increases in the numbers of women and international students in these fields and careers. Unfortunately, many institutions have seen this as sufficient for meeting their diversity goals and have even misclassified some international students and faculty as underrepresented minorities. It should be noted that minority women have not fared as well as white women in the S&E workforce, but they have shown greater increases in degree production. In fact, in 2006, 26 percent of all employed scientists were women. White women represented 69 percent of that total, while minority women represented only 11 percent.

Trends in the participation of women have actually been mixed. In some fields, such as computer science, the participation of women has declined in recent years, and there remains the problem of low percentages of women in STEM faculty in research universities. However, in general, we have achieved greater opportunity for women in some—if not all—fields.

The real story is that of international students. Non-U.S. citizens, particularly those from China and India, have accounted for almost all growth in STEM doctorate awards and, in some engineering fields, have for some time comprised the majority of new doctorate awards. Indeed, temporary

---

[4] Underrepresented minorities, as used in this report, refer to African Americans, Hispanic or Latino Americans, Native Americans and Alaska Natives, and Native Hawaiians and Pacific Islanders. Asians, while a minority group in the U.S. population, are typically overrepresented in science and engineering fields. Pacific Islanders are considered an underrepresented group. However, most national data sets for scientists and engineers aggregate Asians and Pacific Islanders, so it is generally impossible to present separate data for this group. Our focus is, in particular, on students who were born in the United States or who immigrated at an early age and were educated here, rather than individuals who grew up overseas in environments in which they would not be considered a minority and may have benefited from a relatively high quality education.

residents accounted for more than half of the U.S. doctorates in engineering, computer science, and mathematics in 2006. We are coming to understand, then, that relying on the continued growth in the number of non-U.S. citizens in science and technology is an increasingly uncertain proposition, that it does not address our need for more STEM-trained U.S. citizens who are qualified for national security and defense industry positions, that the impending retirements in such fields as geosciences, mathematics, and physics must be a critical concern, and that we must look for other sources of S&E talent for the long run.[5]

For one thing, following the tragic events of September 11, 2001, changes in U.S. visa processing resulted in declines in the numbers of non-U.S. citizens applying for, gaining admission to, and enrolling in graduate study in the United States. Through a series of institutional surveys, the Council of Graduate Schools (CGS) found a substantial decline of 6 percent in first-time international graduate enrollment from 2003 to 2004 and a drop for that period of 3 percent in total graduate enrollment. The next year first-time enrollments increased by 1 percent, but overall enrollment remained down. In subsequent years, the graduate enrollment of international students has increased, but as of 2008, writes CGS, "the rebound in total international enrollment still has not been large enough to reverse the declines that many institutions reported in 2004."[6]

In addition to these data on international enrollment levels, there is cause for concern about whether international students who earn doctorates here will seek to stay and participate in the U.S. science and engineering enterprise or choose to return home or to other parts of the world. An analysis of the percentage of non-U.S. citizen PhDs with temporary visas who earn their degrees from U.S. institutions and then remain in the United States and continue to work found mixed results. The 10-year stay rate in 2007 of those who earned PhDs in 1997 is higher than similarly observed previous 10-year stay rates. However, the five-year stay-rate in 2007 of those who earned PhDs in 2002 is lower.[7] Perhaps more important than these trends, though, is understanding differences in stay rates by country of origin. For example, new doctorates from China, for now, remain in the United States at a very high and fairly stable level over time. Doctorates from India tend to stay at a very high rate but leave over time. Doctorates from Taiwan and South Korea have much lower stay rates and those who initially stay have

[5] National Science Board. *Science and Engineering Indicators.* 2010. Arlington, VA: National Science Foundation.

[6] Council of Graduate Schools. 2008. *Findings from the 2008 CGS International Graduate Admissions Survey, Phase III: Final Offers of Admission and Enrollment.*

[7] Michael G. Finn. 2010. *Stay Rates of Foreign Doctorate Recipients from U.S. Universities, 2007.* Oak Ridge Institute for Science and Education. See http://orise.orau.gov/sep/files/stay-rates-foreign-doctorate-recipients-2007.pdf (accessed February 16, 2010).

a high propensity to leave over time. The key question going forward is whether the stay rates for new doctorates from China will continue as they have in the past or whether these doctorates will begin to return home, as China develops its own higher education sector. There is a very good chance that it will be the latter as China follows the pattern previously set by Taiwan and South Korea.

A Moving Target

If the uncertainty about the future participation of international students suggests a need to ensure that we draw on all demographic sources, the dramatic changes in the demographics of the domestic population suggest that the problem is all the more urgent because the groups that are most underrepresented are also the fastest growing in the population. As shown in Figure 1-1, underrepresented minorities make up 28 percent of the U.S. population but only about 9 percent of the science and engineering workforce. Meanwhile, as shown in Figure 1-2, the U.S. Census Bureau now projects that underrepresented minorities will account for about 45 percent of the U.S. population by the year 2050. So, without a change in course, the current gap between underrepresented minority presence in the population and underrepresented minority participation in S&E will only increase at a time when we most need to close it.

Diversity Is an Asset

Drawing more deeply on diverse groups within our population has benefits beyond meeting the needs for scientists and engineers. Diversity is both a resource for and strength of our society and economy. Scott Page, in *The Difference* (2007), argues that diverse groups are typically smarter and stronger than homogeneous groups when innovation is a critical goal, as it is now in our globally competitive environment.[8] To increase diversity in a population, therefore, strengthens its activity contribution by increasing the number of perspectives and the range of knowledge brought to bear.

There are divergent views among researchers, economists, and others about the costs and benefits of racial and ethnic diversity. Following are examples of these arguments:

- Edwin S. Rubenstein and the National Policy Institute Staff in *The Economic Costs of Racial and Cultural Diversity* (2008):[9] Cultural dif-

---

[8] Scott Page. 2007. *The Difference: How the Power of Diversity Creates Better Groups, Firms, Schools, and Societies.* Woodstock, Oxfordshire, U.K.: Princeton University Press.

[9] E. S. Rubenstein and the Staff of NPI. 2008. *Cost of Diversity: The Economic Costs of Racial and Cultural Diversity*, Issue Number 803, Augusta, GA: National Policy Institute.

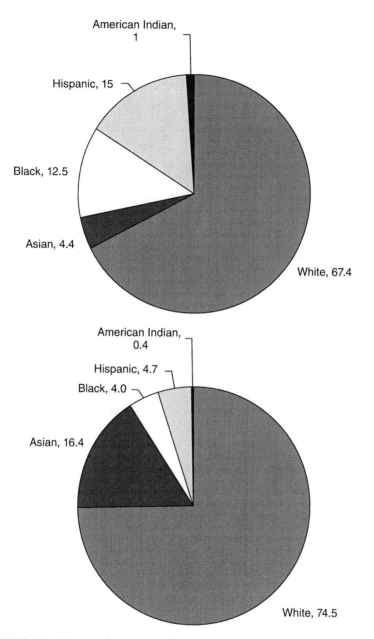

FIGURE 1-1 U.S. population and U. S. science and engineering workforce, by race/ethnicity, 2006.

SOURCE: National Science Foundation, *Women, Minorities, and Persons with Disabilities in Science and Engineering*, Tables A-2 and H-7. Available at http://www.nsf.gov/statistics/wmpd/ (accessed June 12, 2009).

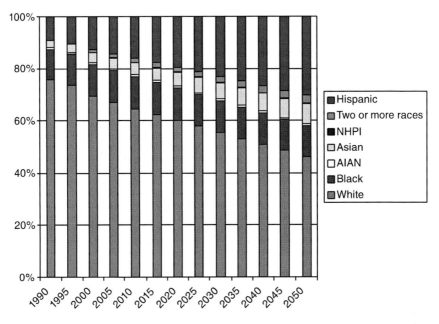

FIGURE 1-2 U.S. population by race/ethnicity, 1990-2050 (2010-2050 projected).
SOURCE: U.S. Census Bureau.

ferences are often a source of social conflict and often act as a barrier to economic progress as well as personal freedom. When societies are multicultural, the ethnocentric differences of race, religion, ethnicity, and language often lead to enmity. Even if different groups live together peacefully, the lack of a common language and common norms reduces cooperation and increases the cost of economic transactions.

• European Commission in *The Costs and Benefits of Diversity* (2003):[10] Companies that implement workforce diversity policies identify important benefits that strengthen long-term competitiveness and, in certain instances, also produce short and medium-term improvements in performance. Companies also face costs of legal compliance, cash costs for additional staff and training, opportunity costs, and business risks.

• Patrick Kelly in *As America Becomes More Diverse: The Impact of State Higher Education Inequality* (2005):[11] Increased educational attainment results in higher personal income, a better-skilled and more adaptable

[10] European Commission. 2003. *The Costs and Benefits of Diversity*. Kent, UK: Centre for Strategy and Evaluation Services.

[11] P. Kelly. 2005. As America Becomes More Diverse: The Impact of State Higher Education Inequality. Boulder, CO: National Center for Higher Education Management Systems.

workforce, fewer demands on social services, higher levels of community involvement, and better decisions regarding healthcare and personal finance. At a time when higher education is increasingly important, some visible race/ethnic groups are consistently in the "have not" category of our society. State policy makers must grasp the social and economic impacts of ignoring the problem.

- Alberto Alesina and Eliana La Ferrara in *Participation in Heterogeneous Communities* (2000):[12] They found that, after controlling for many individual characteristics, participation in social activities is significantly lower in more unequal and in more racially or ethnically fragmented localities.

- Paul Collier in *Ethnicity, Politics and Economic Performance* (2000):[13] Whether diversity affects overall economic growth depends upon the political environment. Diversity is highly damaging to growth in the context of limited political rights, but is not damaging in democracies.

- T. Kochan et al. in *The Effects of Diversity on Business Performance* (2002):[14] Racial diversity has a positive effect on overall performance in companies that use diversity as a resource for innovation and learning. Further, the best performance outcomes occur when diversity is found across entire organizational units.

Several reports present arguments about the impact of diversity in higher education. In *Diversity Works: The Emerging Picture of How Students Benefit,*[15] Daryl G. Smith (1997) concluded that diversity initiatives positively affect both minority and majority students on campus in terms of student attitudes toward racial issues, institutional satisfaction and academic growth. James A. Anderson makes the case in *Driving Change Through Diversity and Globalization* (2008)[16] that the inclusion of diversity and globalization in disciplinary work contributes to the research agendas of individual faculty and their departments, aligns with scholarly values, and promotes such student learning goals as tolerance of ambiguity and paradox, critical thinking, and creativity.

One of the most widely quoted is the study (1999) by Patricia Gurin, professor of psychology and women's study at the University of Michigan.

---

[12] A. Alesina and E. La Ferrara. Participation in heterogeneous communities. *The Quarterly Journal of Economics* 115(3):847-904.

[13] Paul Collier. 2000. Ethnicity politics and economic performance. *Economics and Politics* 12:225-245.

[14] T. Kochan et al. 2002. *The Effects of Diversity on Business Performance*: Report of the Diversity Research Network. Building Opportunities for Leadership Development Initiative, Alfred P. Sloan Foundation and the Society for Human Resource Management.

[15] D. Smith. 1997. *Diversity Works: The Emerging Picture of How Students Benefit.* Washington, DC: Association of American Colleges and Universities.

[16] James A.Anderson. 2008. *Driving Change Through Diversity and Globalization: Transformative Leadership in the Academy.* Sterling, VA: Stylus Publishing.

She presents compelling and comprehensive research that shows the following:[17]

- Structural diversity creates conditions that lead students to experience diversity in ways that would not occur in a more homogeneous student body.
- Students who had experienced the most diversity in classroom settings and in informal interactions with peers showed the greatest engagement in active thinking processes, growth in intellectual engagement and motivation, and growth in intellectual and academic skills.
- The results support the central role of higher education in helping students to become active citizens and participants in a pluralistic democracy. Students who experienced diversity in classroom settings and in informal interactions showed the most engagement in various forms of citizenship and the most engagement with people from different races/cultures.

A preponderance of research suggests that benefits outweigh the various objections to diversity raised in the literature. "Thus, the moral imperative for diversity in higher education is now united with social and economic necessity in a nation that, within a little more than one generation, will be without a racial or ethnic majority. The challenge is to prepare students from all races and backgrounds to work effectively in a decidedly more diverse workplace."[18]

Education Is an Asset

Improving the education of our citizens—especially in science and engineering—has further benefits to society: (1) A citizenry better educated in science and engineering strengthens democracy and informed participation in a world in which STEM is more important than ever to policy; (2) Minority communities will be stronger with greater access to experts who understand science and engineering problems (e.g., water quality and toxic waste dumps) and policy choices for them; and (3) STEM-educated workers will be better able to perform in environments characterized by risk and complexity.

CHARGE TO THE COMMITTEE

Indeed, citing the need to develop a strong and diverse workforce in science, technology, engineering, and mathematics (STEM) fields, U.S. Senators

---

[17] Patricia Gurin, expert report prepared for *Gratz et al. v. Bollinger et al.*, No. 97-75321 (E.D. Mich.) and *Grutter et al. v. Bollinger et al.*, No. 97-75928 (E.D. Mich) (1999).

[18] Frank W. Hale. 2004. *What Makes Racial Diversity Work in Higher Education.* Sterling, VA: Stylus Publishing.

Edward Kennedy, Barbara Mikulski, Patty Murray, and Hillary Clinton, then of the Senate Committee on Health, Education, Labor, and Pensions, wrote to the president of the National Academy of Sciences requesting that the Academy undertake a study that would inform the U.S. Congress about ways to increase underrepresented minority participation in these fields. The U.S Congress later included this study as a mandate in the America COMPETES Act. (A copy of the letter is included in Appendix B.)

The Senators and the COMPETES Act both charged the study committee to explore the role of diversity in the STEM workforce and its value in keeping America innovative and competitive; analyze the rate of change and the challenges the nation currently faces in developing a strong and diverse workforce; and identify best practices and the characteristics of these practices that make them effective and sustainable. They further charged the study committee with addressing the following questions:

1. What are the key social and institutional factors that shape decisions of minority students to commit to education and careers in the STEM fields? What programs have successfully influenced these factors to yield improved results?

2. What are the specific barriers preventing greater minority student participation in the STEM fields? What programs have successfully minimized these barriers?

3. What are the primary focus points for policy intervention to increase the recruitment and retention of underrepresented minorities in America's workforce in the future? Which programs have successfully implemented policies to improve recruitment and retention? Are they "pull" or "push" strategies? Overall, how effective have they been? By what criteria should they be judged?

4. What programs are under way to increase diversity in the STEM fields? Which programs have been shown to be effective? Do they differ by gender within minority group? What factors make them more effective? How can they be expanded and improved in a sustainable way?

5. What is the role of minority-serving institutions in the diversification of America's workforce in these fields? How can that role be supported and strengthened?

6. How can the public and private sectors more effectively assist minority students in their efforts to join America's workforce in these fields?

7. What should be the implementation strategy? The committee should develop a prioritized list of policy and funding action items with milestones and cost estimates that will lead to a science and engineering workforce that mirrors the nation's diverse population.

INFORMATION GATHERING

To carry out this charge, the National Academies appointed a study committee in early 2008. This committee included individuals with expertise in K-12 and higher education, STEM education, STEM employment across sectors, diversity, public policy, and program evaluation. Moreover, committee members represent the range of higher education institutions, from community colleges to research universities. They also include representatives from Historically Black Colleges and Universities (HBCUs), Hispanic-serving Institutions (HSIs), and Tribal Colleges and Universities (TCUs). (See Appendix C for committee member biographies.)

The committee gathered information throughout 2008 through expert testimony, a review of previous reports and the academic literature, and analysis of national data. During three committee meetings (see Appendix D for agendas) on March 10-11, June 11-12, and October 22-23, 2008, the committee heard from the following individuals:

- Charles M. Vest, President, National Academy of Engineering, on innovation and competitiveness policy and the findings and recommendations of the National Academies' report, *Rising Above the Gathering Storm.*
- Representatives from the U.S. Congress, including U.S. Representative Silvestre Reyes, staff from the Offices of U.S Representative Eddie Bernice Johnson and U.S. Representative Michael Honda, and staff from the House Diversity and Innovation Caucus.
- Federal policy and program officials from the White House Office of Science and Technology Policy, the National Science Foundation, the National Institutes of Health, the National Aeronautics and Space Administration, the U.S. Department of Energy, and the U.S. Department of Education.
- Officials from private foundations and programs, including the W.K. Kellogg Foundation, the Howard Hughes Medical Institute, the Association of American Medical Colleges, the UNCF/Gates Millennium Scholars Program, and the Leadership Alliance.
- Experts on the legal and labor market contexts for increasing participation, including Daryl Chubin, AAAS, on *Standing Our Ground: A Guidebook for STEM Educators in the Post-Michigan Era,* and labor economists Mark Regets, National Science Foundation, and Sharon Levin, University of Missouri.
- Experts on demographic trends in STEM fields, including Lisa Frehill, Commission of Professionals in Science and Technology.
- Stakeholder groups, including the National Association of Manufacturers, the National Defense Industry Association, the American Association for the Advancement of Science, the National Action Council for

Minorities in Engineering, and the Society for the Advancement of Chicanos and Native Americans in Science.

• Experts on diversity, mentoring, teacher preparation, K-12 STEM education programs, and minority participation in undergraduate and graduate education, including Shirley Malcom, AAAS.

• Officials from Historically Black Colleges and Universities and other minority-serving institutions (MSIs) on the role of MSIs in broadening participation in STEM fields.

The committee also heard from individuals involved in earlier reports focused on increasing the participation of minorities in STEM fields, including the following:

• Commission on the Advancement of Women and Minorities in Science, Engineering, and Technology Development, *Land of Plenty;*
• National Science and Technology Council, *Ensuring a Strong U.S. Scientific, Technical, and Engineering Workforce in the 21st Century;*
• Building Engineering and Science Talent, *A Bridge for All* and *What It Takes;*
• Willie Pearson Jr., and Diane Martin, *Broadening Participation Through a Comprehensive, Integrated System;*
• National Action Council for Minorities in Engineering, *Confronting the New American Dilemma: Underrepresented Minorities in Engineering;*
• American Association for the Advancement of Science, *In Pursuit of a Diverse Science, Technology, Engineering, and Mathematics Workforce: Recommended Research Priorities to Enhance Participation by Underrepresented Minorities* and other reports; and
• National Research Council, *Assessment of NIH Minority Research Training Programs* and *Understanding Interventions that Encourage Minorities to Pursue Research Careers.*

The committee synthesized this information as a foundation for this report and its findings and recommendations.

ORGANIZATION OF REPORT

This report is organized into three sections. The first section provides an introduction to the issues covered in the report. This chapter provides the context and rationale for the report, as well as a description of the charge to the committee and the committee process. The second chapter in the introductory section presents data to illustrate the dimensions of the problem along the educational pathway and in the science and engineering workforce. The second section of the report, through chapters on preparation, access and motivation, affordability, and academic and social integration, outlines the key educational, social, and professional steps necessary for a student to grow into a scientist or engineer. The paths within the "pathway" or "pipeline" are varied, but elements can be identified to direct discussion of the steps necessary for increasing the participation and success of underrepresented minorities in STEM. The final section of the report consists of two chapters. The first of these chapters provides guiding principles for the development and implementation of policies and programs. The final chapter provides recommendations and a comprehensive list of implementation actions across educational stages and stakeholders. It also includes two priority actions focused on the committee's near-term goal of increasing the persistence and completion of underrepresented minority undergraduates in STEM.

2

Dimensions of the Problem

EDUCATIONAL ATTAINMENT

The connection between education and economic growth in the United States is strong. Claudia Goldin and Lawrence Katz, for example, have argued that it was no coincidence that the twentieth century was both the "American Century," as defined by the growing economic preeminence of the United States, and the "human capital century," as defined by technological change that demanded increasing levels of skill on the part of workers. In the late nineteenth century, they note, technological change in the United States became "skill-biased"—driving demand for an ever more skilled workforce. This skill-demanding technological change was an important force in the United States throughout the twentieth century, with the change brought on by the information technology revolution only the latest chapter, leading to a pattern of increased educational attainment.[1]

Goldin and Katz have summarized that history of increased educational attainment in the United States:

> Not long ago the United States led the world in education and had done so for quite some time. In the 19th century the United States pioneered free and accessible elementary education for most of its citizens. In the early to mid-20th century it extended its lead with the high school movement, when other nations had just discovered mass elementary education. In the

---

[1] C. Goldin and L. F. Katz. 2008. *The Race Between Education and Technology*. Cambridge, MA: The Belknap Press of Harvard University Press, p. 2.

immediate post-World War II era, higher education became a middle-class entitlement in America. A further capstone to the U.S. lead in education in the immediate postwar years was that its universities became the finest in the world. By the 1950s, the United States had achieved preeminence in education at all levels and its triumphant lead would remain undisputed for several decades.[2]

This trajectory in educational attainment was a stunning success and a defining characteristic of both economic growth and our history of social mobility.

Since the 1970s, however, overall educational attainment has stagnated in the United States, even as technological change and the return to higher education—for those who are able to pursue it—have increased. This has happened at the same time as most countries in Europe and several in Asia have caught up and, in some cases, surpassed the United States in educational attainment. Consequently, the United States has lost a key competitive advantage. Once first among OECD nations in postsecondary attainment, the United States has fallen to 11th. In 2008, about 40 percent of 25-to-34-year-olds in the United States had earned a postsecondary degree or credential at the associate's or bachelor's level or above, a level that has not changed significantly in several decades.

Increasing postsecondary success has, as a result, emerged as an important national strategy and goal for ensuring a strong workforce and competitive economy for the future. The College Board has urged that we increase the percentage of the 25- to 34-year age group with postsecondary degrees (associate, baccalaureate, or above) to 55 percent.[3] The Lumina Foundation has adopted a goal, through its Making Opportunity Affordable program, to "raise the proportion of the U.S. adult population who earn college degrees to 60 percent by the year 2025, an increase of 16 million graduates above current rates" (2008).[4] President Obama (2009) has challenged the United States to have the highest proportion of postsecondary graduates in the world by 2020.[5]

Patterns of racial participation in education overlay this history in a critical way. Underrepresented minorities were largely and systematically excluded from mainstream educational opportunities through de jure and de facto segregation that continued from *Plessy v. Ferguson* in 1896 through the desegregation and busing battles of the 1970s. This period of exclusion

---

[2] Goldin and Katz. 2008. *Race Between Education and Technology*, p. 324.

[3] The College Board. 2009. *Coming to Our Senses: Education and the American Future.*

[4] The Lumina Foundation, http://www.luminafoundation.org/our_work/ (accessed March 27, 2009).

[5] President Barack Obama, Address to Joint Session of Congress, February 24, 2009. http://www.whitehouse.gov/the_press_office/remarks-of-president-barack-obama-address-to-joint-session-of-congress/ (accessed September 4, 2009).

coincides with the period of increasing educational opportunity for white Americans discussed above. The efforts of the civil rights movement led to increases in educational opportunity for underrepresented minorities, beginning in the 1940s with *Mendez et al. v. Westminster Schools District of Orange County,* continuing in the 1950s with the landmark *Brown v. Board of Education of Topeka,* and accelerating in the 1960s, 1970s, and after with cases such as *Edgewood ISD v. Kirby.*

This period of inclusion for underrepresented minorities, however, particularly from the 1970s on, coincides with stagnation in both public educational investment and overall levels of educational attainment. So, little progress has been made to more than marginally improve educational outcomes for minorities.[6] While the targeted level of 55 percent postsecondary attainment is already achieved by Asian Americans in the United States and nearly matched by our white population (as it is by their peer cohorts in Canada and Japan), the postsecondary attainment of underrepresented minority students lags behind that of white and Asian students dramatically. Underrepresented minorities will need to more than *double* their proportions with a postsecondary degree in order just to meet the 55 percent mark. At present, just 26 percent of African Americans, 24 percent of Native Americans and Pacific Islanders, and 18 percent of Hispanics and Latinos in the 25- to 34-year-old cohort have attained at least an associate's degree.

The news is even worse in science, technology, engineering, and mathematics (STEM) fields, the subject of this report. In 2000, the United States ranked 20 out of 24 countries in the percentage of 24-year-olds who had earned a first degree in the natural sciences or engineering, and *Rising Above the Gathering Storm* recommended efforts to increase the percentage of 24-year-olds with these degrees from 6 percent to at least 10 percent, the benchmark already attained by Finland, France, Taiwan, South Korea, and the United Kingdom.

But again, as bad as the statistics are for the overall population, they are even more alarming for underrepresented minorities. These students now need to triple, quadruple, or even quintuple their proportions with a first degree in these fields in order to achieve this 10 percent goal. At present, just 2.7 percent of African Americans, 3.3 percent of Native Americans and Alaska Natives, and 2.2 percent of Hispanics and Latinos who are 24 years old have earned a first degree in the natural sciences or engineering.[7]

---

[6] C. Newfield. 2008. *Unmaking the Public University: The Forty-Year Assault on the Middle Class.*(Cambridge, MA: Harvard University Press.

[7] National Science Foundation, *Women, Minorities, and Persons with Disabilities in Science and Engineering,* http://www.nsf.gov/statistics/wmpd/ (accessed March 27, 2009); U.S. Census Bureau, Population Estimates, Available at http://www.census.gov/popest/national/asrh/NC-EST2007-asrh.html (accessed March 27, 2009).

The national goal of increased postsecondary educational attainment is vital. The goal of increased postsecondary participation and success for underrepresented minorities in STEM, which relies in part on the former goal, is strategically important and, as we have now seen, a task of formidable scale.

EDUCATION AND THE
SCIENCE AND ENGINEERING WORKFORCE

The S&E workforce is large and fast-growing: more than 5 million strong and projected by the U.S. Bureau of Labor Statistics to grow faster than any other sector in coming years.[8] This growth rate provides an opportunity to draw on new sources of talent, including underrepresented minorities, to make our S&E workforce as robust and dynamic as possible.

The data on underrepresented minorities in the S&E workforce, however, suggest that while there has been needed progress, there is also reason for continued concern, even alarm. For example, the percentage of our college-educated, nonacademic S&E labor force that is African American increased from 2.6 percent in 1980 to 5.1 percent in 2005, and the percentage that is Hispanic increased from 2.0 percent to 5.2 percent during that period.[9] However, these percentages and the progress they represent remain small and insufficient, as African Americans comprise 11 percent and Hispanics 14 percent of the U.S. civilian labor force, and even higher percentages in the U.S. population.

Indeed, the proportion of underrepresented minorities in S&E would need to *triple* to match their share of the overall U.S. population, revealing a scale of effort that is substantial. As Figure 2-1 shows, in 2006 underrepresented minority groups represented 28.5 percent of our national population but just 9.1 percent of college-educated Americans in science and engineering occupations (academic and nonacademic). Data show that in 2006, fewer than 10 percent of STEM faculty at research universities were

---

[8] Out of a civilian labor force of more than 150 million in the United States, the S&E workforce ranges in size from less than 4 million to more than 21 million, depending on definitions used, such as occupation, field of degree, and the extent to which S&E knowledge is needed for employment. Here we focus on the most commonly used definition of the S&E workforce, namely, those individuals with a bachelor's degree or above working in an S&E occupation.

U.S. Department of Labor, Bureau of Labor Statistics, Current Employment Statistics, Employment Situation, Table A-1, Employment status of the civilian population by sex and age, http://www.bls.gov/news.release/empsit.t01.htm (accessed June 16, 2009). National Science Board, *Science and Engineering Indicators, 2008*, 3-8; and Sidebar, "Who Is a Scientist or Engineer?," pp. 3-9.

[9] Table underlying Figure 3-27 in *Science and Engineering Indicators, 2008*.

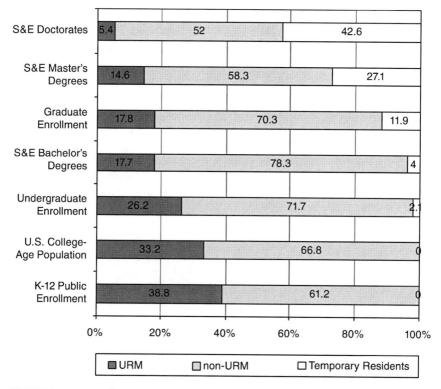

FIGURE 2-1 Enrollment and degrees, by educational level and race/ethnicity/citizenship, 2007.
SOURCES: NCES, Digest of Education Statistics, 2008, Table 41. NSF, Women, Minorities, and Persons with Disabilities, Tables A-2, C-6, E-3, and F-11. NSF, S&E Degree Awards, 2006, Table 3.

underrepresented minorities; the percentage of URM women is even lower.[10] Whites were overrepresented at 74.5 percent of the S&E workforce compared to 67.4 percent of the U.S. population. Asians were overrepresented as well: The proportion of Asians in the S&E workforce (16.4 percent) is substantially more than their representation in the U.S population (4.4 percent).

Underrepresentation of this magnitude in the S&E workforce stems from the underproduction of minorities in S&E at every level of post-secondary education, with a progressive loss of representation as we proceed

---

[10] D. Nelson. 2007. *A National Analysis of Minorities in Science and Engineering Faculties at Research Universities*. Norman, OK: Diversity in Science Association and University of Oklahoma.

up the academic ladder. In 2007, as shown in Figure 2-1, underrepresented minorities made up 38.8 percent of K-12 public enrollment, 33.2 percent of the U.S college age population, 26.2 percent of undergraduate enrollment, and 17.7 percent of those earning science and engineering bachelor's degrees. In graduate school, underrepresented minorities comprise 17.7 percent of overall enrollment but are awarded just 14.6 percent of S&E master's degrees and a miniscule 5.4 percent of S&E doctorates.

These trends are seen in each underrepresented racial/ethnic group:

• In 2006, **Hispanic or Latino Americans** comprised 15.0 percent of the U.S. population and 17.8 percent of the college-age population, age 18-24. However, in 2005, they earned 7.9 percent of S&E bachelor's degrees and 6.2 percent of S&E master's degrees. In 2007, they earned 5.2 percent of S&E doctoral degrees awarded by U.S institutions to U.S. citizens and permanent residents and just 2.9 percent of S&E doctorates awarded to all recipients (including non-U.S. citizens who are temporary visa holders).

• In 2006, **African Americans** comprised 12.5 percent of the U.S. population and 14.1 percent of the college-age population, age 18-24. However, in 2005, they earned 8.8 percent of S&E bachelor's degrees and 8.8 percent of S&E master's degrees. In 2007, they earned 4.5 percent of S&E doctoral degrees awarded by U.S institutions to U.S. citizens and permanent residents and just 2.5 percent of S&E doctorates awarded to all recipients (including non-U.S. citizens who are temporary visa holders).

• In 2004, **Native Americans and Alaska Natives** comprised 0.8 percent of the U.S. population and 1.0 percent of the college-age population, age 18-24. In 2005, they earned 0.7 percent of S&E bachelor's degrees and 0.6 percent of S&E master's degrees. In 2007, they earned 0.5 percent of S&E doctoral degrees awarded by U.S institutions to U.S. citizens and permanent residents and just 0.3 percent of S&E doctorates awarded to all recipients (including non-US citizens who are temporary visa holders).

All of these indicators point to underutilization in science and engineering fields of persons from these minority groups, with especially severe underproduction at the doctoral level.

TRACKING POSTSECONDARY INTEREST AND COMPLETION

Research on underproduction of minorities in science and engineering has focused on interest and persistence. In 2005, the American Council on Education (ACE), analyzing data from the 1990s collected by the National Center for Education Statistics, found that although the proportion of African American and Hispanic students who begin college with an interest in majoring in STEM was similar to the proportion of white and Asian American students, African American and Hispanic students completed STEM degrees after six years at a lower rate.[11] In particular, ACE found:

• African American and Hispanic students begin college interested in majoring in science, technology, engineering, and mathematics (STEM) fields at rates similar to those of white and Asian American students: In the 1995-1996 academic year, 18.6 percent of African American students and 22.7 percent of Hispanic students began college interested in majoring in STEM fields compared with 44.4 percent of white and Asian-American students.
• African American and Hispanic students persist in these fields through their third year of study. By the spring of 1998, students in each racial/ethnic group continued to study STEM fields at nearly the same rates (56 percent of African Americans and Hispanics, 57 percent of whites and Asian Americans).
• African American and Hispanic students did not earn their bachelor's degrees at the same rate as their peers. By the spring of 2001, 62.5 percent of African Americans and Hispanics majoring in STEM fields had completed a bachelor's degree compared with 94.8 percent of Asian Americans and 86.7 percent of whites.

These findings are important, yet they are based on a cohort of students that began college almost 15 years ago.

Recently, the Higher Education Research Institute (HERI) at the University of California Los Angeles released data from a sample of more than 200,000 students across 326 four-year institutions that began college in fall 2004, providing trends in aspirations to major and completion of degrees in STEM disaggregated by race/ethnicity. The data allow us to examine current trends and see whether there has been substantial change from the mid-1990s.[12]

As shown in Figure 2-2, HERI found that while there has been considerable volatility in aspiration to major in STEM since 1971, trends in aspiration by race/ethnicity began to converge in the late 1980s and have stabilized at between 30 and 35 percent both overall and for white/Asian American and underrepresented minority groups since the early 1990s. While the percentages aspiring to a STEM major are higher in the HERI

[11] American Council on Education. 2005. *Increasing the Success of Minority Students in Science and Technology.* Washington, DC: ACE.
[12] Higher Education Research Institute at UCLA, *Degrees of Success: Bachelor's Degree Completion Rates Among Initial STEM Majors,* HERI Report Brief, January 2010. http://www.heri.ucla.edu/nih/HERI_ResearchBrief_OL_2010_STEM.pdf (accessed February 20, 2010).

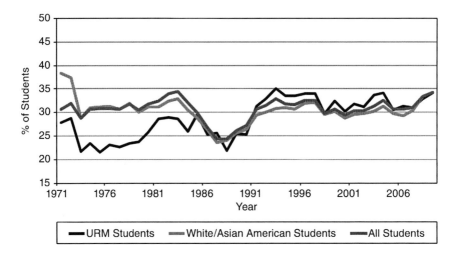

FIGURE 2-2 Trends in students' aspiration to major in a STEM discipline by racial identification, 1971-2009.
SOURCE: University of California Los Angeles, Higher Education Research Institute.

data than in the NCES data (which was based on a much smaller sample), the overall finding is the same: Underrepresented minorities report a level of aspiration to major in STEM similar to those of their white/Asian peers.

As shown in Figure 2-3, HERI examined four-year (2008) and five-year (2009) completion rates of the 2004 STEM majors by race/ethnicity, finding that underrepresented minorities completed at a much lower rate at both intervals relative to their white and Asian American peers. White and Asian American students who started as STEM majors have four-year STEM degree completion rates of 24.5 and 32.4 percent, respectively. In comparison, Latino, black, and Native American students who initially began college as STEM majors had four-year STEM degree completion rates of 15.9, 13.2, and 14.0 percent, respectively. As HERI reports, the differences after 5 years is even more pronounced. Approximately 33 and 42 percent of white and Asian American STEM majors, respectively, completed their bachelor's degree within 5 years of college entry. In contrast, the five-year completion rates for Latino, black, and Native American students were 22.1, 18.4, and 18.8 percent, respectively.

HERI data show four- and five-year completion rates for the 2004 cohort (the six-year completion rate will be available later this year), and the NCES data analyzed by ACE provide a six-year completion rate. However, the gaps in STEM completion rates of STEM majors between underrepre-

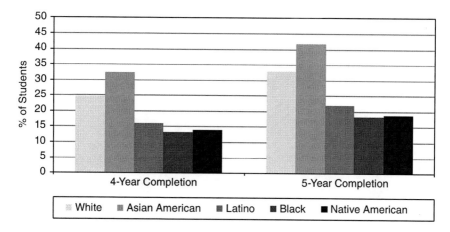

FIGURE 2-3 Percentage of 2004 STEM aspirants who completed STEM degrees in four and five years, by race/ethnicity.
SOURCE: University of California Los Angeles, Higher Education Research Institute.

sented minorities and whites and Asian Americans are similarly large for the 1995 and 2004 cohorts, and, in the case of the 2004 cohort, the gap appears to increase as the interval from matriculation grows.

Another salient dimension to the picture of STEM completion for underrepresented minorities is the difference in completion rates for under-represented minorities in STEM relative to those for underrepresented minorities who major in non-STEM fields. As shown in Figure 2-4, all five racial/ethnic groups have higher four- and five-year completion rates in non-STEM majors. This analysis reveals a trend that is relevant to both whites and Asian Americans as well as underrepresented minorities. That is, there is a problem for STEM completion relative to non-STEM completion as well as a problem for underrepresented minorities in STEM relative to their white and Asian American peers.

After further analyzing the NCES data, ACE identified several key differences in the data between students who earned a bachelor's degree by spring 2001 in a STEM field and those who did not (noting that there may be other differences that had not been counted in the data).

- Completers were better prepared for postsecondary education because a larger percentage took a highly rigorous high school curriculum.

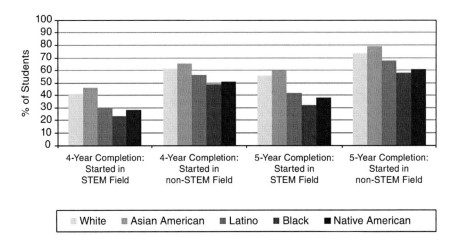

FIGURE 2-4 Four- and five-year degree completion rates of 2004 freshmen, by initial major aspiration and race/ethnicity.
SOURCE: University of California Los Angeles, Higher Education Research Institute.

• Nearly all completers were younger than 19 when they entered college in 1995-1996 compared with 83.9 percent of noncompleters.
• Completers were more likely to have at least one parent with a bachelor's degree or higher.
• Completers came from families with higher incomes.
• Noncompleters were more likely to work 15 hours or more a week.

That is, *preparation, motivation,* and *financial support* are important to success and completion. Moreover, all of these can be the focus of immediate intervention.

HERI has not yet released an analysis of the differences between completers and noncompleters, but we can expect that there will be an overlap in the issues that have been at play for the 2004 HERI cohort as ACE found for the 1995 NCES cohort. Indeed, based on research that will be discussed below, we would expand this list of factors affecting completion to include preparation, access to information, self-motivation and identification with science or engineering as a profession, institutional strategies for inclusion, and professional development. ACE also found that "strategies for increasing the degree completion of minority students in the STEM fields are the same for increasing success in any other major," a conclusion similar to that of Daryl Chubin and Wanda Ward, who have examined features of programs designed to increase participation of underrepresented minorities in STEM.[13] However, a particular problem appears to exist for STEM programs, as evidenced by the HERI completion data.

TAKING STOCK

Trends in the overall number of underrepresented minorities earning science and engineering degrees are encouraging. However, just as the persistence data we have examined confirms that there remains a problem in persistence and completion for underrepresented minorities relative to their white and Asian American peers, so too do data on the relative proportions of each racial/ethnic group among those earning science and engineering degrees.

Science and Engineering Degrees

Indeed, we have achieved important progress in increasing the participation of underrepresented minorities in higher education generally and in science and engineering specifically. For example, there was a 77 percent increase in S&E associate's degrees awarded to underrepresented minorities from 1998 to 2007, with an increase of about 50 percent in computer sciences.[14] Community colleges face the same challenges in retaining students as do other institutions, or even more than they do. Many incoming freshmen lack the basic mathematics and science prerequisites for persistence, especially in urban communities that serve a large minority population from low-performing high schools, and the institutions are forced to provide intensive programs in remedial education to increase minority student retention in STEM. Chang (2003) noted that community colleges have implemented innovative approaches to retain underrepresented students.[15] Some institutions now offer programs that provide students an opportunity to engage in hands-on projects, and others have changed the curriculum to promote more collaborative group work. According to Chang, these "social support systems are of particular benefit to underrepresented minorities in fields that have previously been perceived as intimidating or unwelcoming." Community colleges also are seeking to increase the admission and transfer of underrepresented minorities through partnerships with elementary and

---

[13] Daryl E. Chubin and Wanda E. Ward. Building on the BEST principles and evidence: A framework for broadening participation, in M. Boyd and J. Wesermann, eds., *Broadening Participation in Undergraduate Research: Fostering Excellence and Enhancing the Impact.* Washington, DC: Council of Undergraduate Research, forthcoming.

[14] National Science Foundation. 2009. *Women, Minorities, and Persons with Disabilities in Science and Engineering.* The totals exclude associate's degrees in psychology and social sciences.

[15] J. C. Chang. 2003. Women and minorities in the science, mathematics, and engineering pipeline, *ERIC Digest.*

TABLE 2-1 Percentage Change in S&E Degrees Earned, by Degree Level and Race/Ethnicity (Bachelor's and Master's 1998-2007; Doctorates 1998-2007)

	Percent Change		
	Bachelor's	Master's	Doctorate
White	14.7	11.9	−2.3
Asian/Pacific Islander	33.8	38.6	30.1
Black	31.3	70.9	44.3
Hispanic	49.9	74.4	62.3
American Indian/Alaska Native	38.8	48.8	3.9
Other/Unknown	165.5	149.4	69.0
Temporary Residents	17.8	26.8	50.4

SOURCE: National Science Foundation. 2009. *Women, Minorities, and Persons with Disabilities in Science and Engineering.*

secondary schools and four-year institutions. Thus, the community college, with its diverse student population, is an integral player in advancing minority representation in STEM.

Meanwhile, as shown in Table 2-1, underrepresented minorities also are the fastest growing populations in science and engineering at the bachelor's and master's levels, as indicated by degrees awarded by four-year institutions. Their numbers are growing faster than those of temporary residents and whites and are outstripped only by the other/unknown category.[16]

However, and this cannot be stressed enough, *this progress is comprised of large gains over a very small base,* and minorities remain underrepresented across science and engineering fields and academic levels. Indeed, representation varies across fields, with some showing trivial progress and representation decreases as we ascend the academic ladder.

As shown in Figure 2-5, there is considerable variation in representation across S&E fields. At the **bachelor's level,** there is strongest representation in biological sciences, computer science, social sciences, and psychology. In other fields, though, there are much smaller levels of participation, especially in astronomy, materials engineering, and earth, atmospheric, and ocean sciences. At the **master's level,** representation is generally lower across all fields, though the pattern of representation is similar. It is strongest in industrial engineering, the social sciences, and psychology. Representation at the **doctoral level** is the most problematic and should be the focus of significant intervention.

---

[16] This latter category is largely composed of individuals who report multiple races or refuse to respond to race/ethnicity questions. Individuals of underrepresented minority ancestry comprise the majority of these groups, so the growth of this category may result in some level of underreporting of minority participation. NSF/SRS documentation.

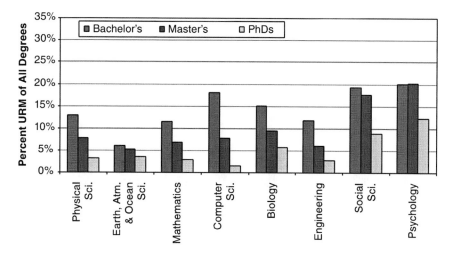

FIGURE 2-5 Underrepresented minorities among S&E degree recipients, by degree level, 2006.
SOURCE: Commission on Professionals in Science and Technology.

Table 2-1, again, shows that the proportion of underrepresented minorities has recently increased for all groups. Hispanics earning S&E doctorates increased more than 66 percent from 1998 to 2007. African Americans made more modest gains of 44.3 percent during that period. (In both cases, again, these are gains over a very small base.) The increases among Hispanics and African Americans partially compensated for decreases during this period in the numbers of whites and Asians earning S&E doctorates (the downward trend in doctoral degrees awarded to whites and Asians turned around in 2003 and are heading back to pre-2000 levels). However, playing an even larger role are non-U.S. citizens on temporary visas.

There is considerable variation in underrepresented minority participation by field at the doctoral level. As also seen in Figure 2-5, underrepresented minorities comprise extremely low percentages in the natural sciences and engineering—biology at 6 percent, the physical sciences and engineering below 5 percent—and numbers so low in computer science as to make them practically nonexistent. Representation is highest for these groups, again, in the social sciences and psychology. However, there is variation in representation within these latter fields. For example, within sociology, psychology, economics, and political science, African Americans tend to be substantially underrepresented in quantitative subfields, such as statistics, sociology of science, psychometrics, and econometrics.

As shown in Figure 2-6, it is in the fields where underrepresented minorities have extremely low representation that we find the highest levels

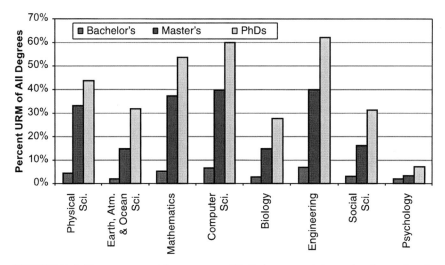

FIGURE 2-6 Temporary residents among S&E degree recipients, by degree level, 2006.
SOURCE: Commission on Professionals in Science and Technology.

of non-U.S. citizens with temporary visas. In 2007, 60 percent or more of doctorates awarded by U.S. institutions in engineering and computer science were to temporary visa holders. High percentages were also awarded to this group in mathematics and the physical sciences. Temporary visa holders also received moderately high percentages of doctoral awards in earth, atmospheric, and ocean sciences; biology; and the social sciences (chiefly in economics). By contrast, their awards in psychology—one field with relatively higher awards to underrepresented minorities—are very low.

From 1998 to 2007, temporary visa holders increased their numbers in S&E doctorate awards by 50.4 percent and were, therefore, one of the fastest growing groups by far. This increase continued over time during the post-September 11 period when there was significant concern about the application, acceptance, and enrollment of non-U.S. citizens at the graduate level. We have yet to see the effect of these post-911 trends as most of those earning doctorates during the 1998-2007 period began their studies before 2001. Based on trends in graduate enrollment for this group, we might assume that there will be decreases in S&E doctorates among them in the near future.

Doctoral Workforce

The doctoral workforce is of particular importance and interest. Not only does it provide underrepresented minorities an opportunity to contribute to teaching and research, but it is at this level that increases can also have a multiplier effect. It becomes the pool for higher education institutions to recruit and develop the talent to diversify their faculty. However, diversifying faculties is perhaps the least successful of the diversity initiatives for a number of reasons cited in higher education research, such as unwelcoming climates at predominantly white institutions (Turner and Myers, 2000), inequity in hiring and promotion practices (Rowan-Kenyon and Milem 2008), and the presumption that minorities who do not earn their degrees at the most prestigious institutions are less qualified (Mickelson and Oliver, 1991). As the number of underrepresented minorities in faculty positions increases, the more role models underrepresented minority students have who "look like them" and the higher the rate at which underrepresented minority students enroll and graduate. Three African American chemists, for example, are responsible for mentoring close to 400 minority students in the field who then went on to earn PhDs and, for the most part, to enter academic careers.[17]

However, the level of underrepresented minority participation in the doctoral S&E workforce is very small. As shown in Figure 2-7, underrepresented minorities as a whole comprised just 8 percent of academic doctoral scientists and engineers working in four-year colleges and universities in 2006. The percentage of doctorate holders in nonacademic S&E occupations who are underrepresented minorities increased from 4.4 percent in 1990 to 6.1 in 2005, a substantial increase if it were not over a very small base.[18] Myers and Turner concluded that market forces such as wages play a more prominent role in affecting faculty representation in the short run than pipeline factors designed to increase the supply of minority faculty.[19]

Overall, underrepresented minorities comprise just 6.8 percent of doctoral scientists, and there is even worse news about their participation in high-end research. As shown in Table 2-2, data from the National Institutes of Health show that African Americans and Hispanics are even more underrepresented among their principal investigators (PIs). In 2006, only 1.8 percent of PIs receiving NIH research grants were African Americans and only 3.5 percent were Hispanic. Similarly, as shown in Table 2-3, 2.2 percent of PIs awarded NSF research grants were African Americans, 4.0 percent were Hispanic, and 0.3 percent were Native American/Alaska Native/Native Hawaiian/Pacific Islander.

---

[17] Isiah Warner. *A Tale of Three Chemists*. Presentation to Study Committee, Third Committee Meeting, October 22, 2008.

[18] National Science Board, *Science and Engineering Indicators, 2008*, Table underlying Figures 3-28.

[19] S. Myers Jr. and C. Turner. The effects of Ph.D. supply on minority faculty representation, *The American Economic Review* 94(2), Papers and Proceedings of the One Hundred Sixteenth Annual Meeting of the American Economic Association. San Diego, CA, pp. 296-301.

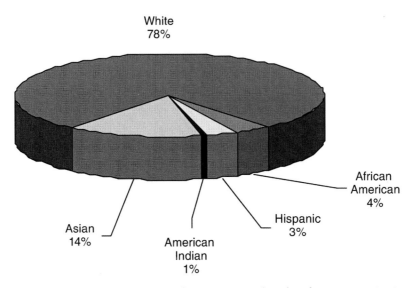

FIGURE 2-7 Doctoral scientists and engineers employed in four-year institutions, by race/ethnicity, 2006.
SOURCE: Commission on Professionals in Science and Technology.

TABLE 2-2 Principal Investigators on NIH Research Grants, by Race/Ethnicity

Fiscal Year	White	African Americans[a]	Hispanic[b]	Other[c]
2000	86.2%	1.3%	2.9%	11.4%
2001	85.7%	1.3%	2.9%	12.1%
2002	85.2%	1.5%	3.1%	12.4%
2003	84.4%	1.6%	3.3%	13.2%
2004	83.5%	1.7%	3.3%	14.1%
2005	82.8%	1.7%	3.5%	14.8%
2006	82.1%	1.8%	3.5%	15.4%

[a] Race data may contain individuals reporting Hispanic ethnicity, as well as individuals reporting more than one race

[b] "All Hispanic" includes Hispanic Race, plus individuals reporting Hispanic Ethnicity (for these individuals the data includes individuals who are represented in one or more of the racial groups)

[c] Includes Asian, Native Hawaiian or Pacific Islander, and American Indian or Alaska Native.

SOURCE: Raynard Kington, Deputy Director, National Institutes of Health, Presentation to Committee, June 11, 2008.

TABLE 2-3 NSF Research Proposals and Awards, by Race/Ethnicity of PI, 2009

American Indian/	Proposals	78	% of Proposals	0.2%
Alaska Native	Awards	27	% of Funded	0.2%
	Funding Rate	35%		
Black/African American	Proposals	1,005	% of Proposals	2.5%
	Awards	290	% of Funded	2.2%
	Funding Rate	29%		
Hispanic or Latino	Proposals	1,724	% of Proposals	4.2%
	Awards	529	% of Funded	4.0%
	Funding Rate	31%		
Native Hawaiian/	Proposals	20	% of Proposals	0.0%
Pacific Island	Awards	8	% of Funded	0.1%
	Funding Rate	40%		
Asian	Proposals	9,377	% of Proposals	23.1%
	Awards	2,426	% of Funded	18.2%
	Funding Rate	26%		
White, Not of	Proposals	28,476	% of Proposals	70.0%
Hispanic Origin	Awards	10,023	% of Funded	75.3%
	Funding Rate	35%		
All Races	Proposals	40,680	% of Proposals	100.0%
	Awards	13,303	% of Funded	100.0%
	Funding Rate	39%		

SOURCE: National Science Foundation, NSF Enterprise Information System, as of October 1, 2009.

In sum, underrepresented minorities are underutilized in science and engineering. There is underproduction of S&E graduates at every educational level from secondary school through doctoral education. Underrepresented minorities are also significantly underrepresented in the doctoral population, in the faculty, and among researchers awarded federal research funds. This is a substantial human resource for the United States in general and United States science and engineering in particular, and we are currently squandering it.

BECOMING SCIENTISTS
AND ENGINEERS

3

Preparation

FROM A NATION AT RISK TO AMERICA AT THE CROSSROADS: K-12

More than 25 years ago, the National Commission on Excellence in Education issued the landmark report, *A Nation at Risk: The Imperative for Educational Reform*. This report argued that the nation's education system was "being eroded by a rising tide of mediocrity that threatens our very future as a nation and as a people." Academic achievement test scores were falling; fewer students were adequately prepared for entry into college or the job market; and schools were failing to compete with those in other developed countries.

Later that year, the National Science Board Commission on Precollege Education in Mathematics, Science and Technology published the report *Educating Americans for the 21st Century*, responding to the impact of emerging new technologies on K-12 education. These reports occurred during a time when the demand for highly skilled workers in emerging fields was accelerating rapidly.

They called for massive reform in the educational process "at the expense of a strong public commitment to the equitable treatment of our diverse population." Subsequently, former President George H. W. Bush convened a historic Education Summit at Charlottesville, Virginia, in 1989 with 50 governors at which they agreed to set national education goals. The Bush administration and the governors announced the six national

All, regardless of race or class or economic status, are entitled to a fair chance and to the tools for developing their individual powers of mind and spirit to the utmost. This promise means that all children by virtue of their own efforts, competently guided, can hope to attain the mature and informed judgment needed to secure gainful employment, and to manage their own lives, thereby serving not only their own interests but also the progress of society itself.

– A Nation At Risk, April 1983

education goals[1] (Box 3-1) and created the National Education Goals Panel[2] to report national and state progress toward the goals, identify promising practices for improving education, and help to build a nationwide bipartisan consensus to achieve the goals. The Goals Panel released annual reports and other resource documents as guidance for measuring progress toward the goals, establishing national education standards, assessing students' completion of school, and recognizing the link between teacher quality and student achievement.

The No Child Left Behind Act (NCLB) of 2001 that pushed for increased accountability for states, school districts, and schools; more choices for parents and students, especially those attending low-performing schools; greater flexibility for states and school districts in the use of federal education funds in exchange for improved performance; and a stronger emphasis on reading. Tough sanctions would be imposed on schools failing to show improved performance, and those that narrowed the achievement gaps would be eligible to receive State Academic Achievement Awards. The principles of the NCLB Act also flowed to other programs authorized by the Elementary and Secondary Education Act of 1965, such as the Improving Teacher Quality State Grants program that applies scientifically based research to prepare, train, and recruit high-quality teachers. More recently, under President Barack Obama, the American Recovery and Reinvestment Act of 2009 provided $4.35 billion for the Race to the Top Fund, a competitive grant program designed to encourage and reward states that are creating the conditions for education innovation and reform.

In spite of the numerous reports and policy and reform initiatives targeting curriculum and educational standards, assessments, and teacher preparation, today the nation is faced with the same issues—failing schools and

---

[1] The six goals were later expanded to eight by Congress.

[2] The Goals Panel was reconstituted to include representatives from Congress as voting members and equal numbers of Republicans and Democrats. President Clinton signed the "Goals 2000: Educate America Act" adding state legislators to the panel membership.

BOX 3-1
Education Goals 2000

The 1989 Education Summit led to the adoption of six National Education Goals, later expanded to eight by Congress. Essentially, the goals state that by Year 2000:

1. All children will start school ready to learn.
2. The high school graduation rate will increase to at least 90%.
3. All students will become competent in challenging subject matter.
4. Teachers will have the knowledge and skills that they need.
5. U.S. students will be first in the world in mathematics and science achievement.
6. Every adult American will be literate.
7. Schools will be safe, disciplined, and free of guns, drugs, and alcohol.
8. Schools will promote parental involvement and participation.

SOURCE: Goals 2000—The Clinton Administration Education Program, http://www.nd.edu/~rbarger/www7/goals200.html.

---

inequitable education at a time when there is even more need for a skilled workforce. Recent reports show that previous efforts have produced mixed results for the general populace and have had limited effectiveness in bridging the achievement gap for underrepresented minorities, the fastest growing segment of the U.S. population. In fact, the efforts have failed to address the special needs of underrepresented minorities in a fashion systematic enough to sustain the small gains made. The problem has been exacerbated by a surge in the nation's Hispanic population due to substantial immigration since the 1990s that has filled many schools with large numbers of children who are not native speakers of English. Thus, as underrepresented minorities continue to be unprepared to matriculate successfully through the education trajectory, the United States continues to fall further behind other industrialized nations in academic achievement and degree production in science and engineering.

NATIONAL MARKERS FOR UNDERREPRESENTED MINORITIES

A range of indicators signal the need for us to reconsider the efficacy of national policies and investments in K-12 education. These are presented in the context of the demographic shifts in the American population and the potential impact of continuing the legacy of inequality in the educational system. There are systemic failures in the implementation of federal, state,

and local policies designed to provide equity and excellence in K-12 education, and these failures weaken our foundation for future prosperity.

K-12 Enrollment Trends

According to *Projections of Education Statistics to 2018,* total public
and private elementary and secondary school enrollment reached a record
55 million in fall 2006 and is projected to set new records each year
from 2009 through 2018, with increasing proportions of underrepresented
minorities. The South is expected to maintain the largest overall enrollment, with 40 percent of students residing in this region. Private school
enrollment is expected to decrease during this period, given its 9 percent
enrollment growth between 1985 and 2008 compared to the 26 percent
growth in public schools.[3]

The proportion of underrepresented minorities enrolled in public elementary and secondary schools has increased over time. Figure 3-1 shows
that between 1972 and 2007, the percentage of public school students
who were white decreased from 78 to 56 percent, while the percentage of
students from other racial/ethnic groups increased from 22 to 44 percent,
largely reflecting the growth in the percentage of Hispanic students

Thus, the K-12 pipeline is expected to have an inevitable majority of
underrepresented minorities and must be a major focal point of intervention
to cultivate the diverse talent pool needed to sustain the nation's future in
STEM. The K-12 pipeline can be divided into four key transition points
for the purposes of policy intervention for underrepresented minorities:
prekindergarten, elementary school, middle school, and high school. There
are indicators for each of these transition points that signal the need for
intervention and that impact the continuing progression of underrepresented
minority students.

International Comparisons of K-12 Mathematics and Science Performance

International comparisons provide a window through which to view
our nation's competitiveness in the global economy. These comparisons spur
a review of policy issues from access to education to equity of resources
devoted to educational achievement, and they point to the need for more
effective and coherent strategies to improve academic performance.

For example, the 2007 Trends in International Mathematics and Science
Study (TIMSS) reports that math and science scores for U.S. 4th and 8th
grade students were lower than those of students in peer countries, accord-

---

[3] W. J. Hussar and Bailey, T. M. 2009. Projections of Education Statistics to 2018 (NCES
2009-062). National Center for Education Statistics, Institute of Education Sciences, U.S.
Department of Education, Washington, DC.

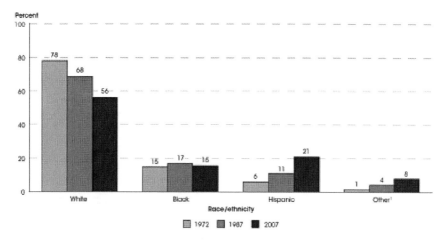

FIGURE 3-1 Percentage distribution of public school students enrolled in kindergarten through 12th grade by race/ethnicity: Selected years, October 1972-October 2007.

NOTE: "Other" includes all students who identified themselves as being Asian, Hawaiian, American Indian, or two or more races. Estimates include all public school students enrolled in kindergarten through 12th grade. Race categories exclude persons of Hispanic ethnicity. Over time, the Current Population Survey (CPS) has had different response options for race/ethnicity.

SOURCE: U.S. Department of Commerce, Census Bureau, Current Population Survey (CPS), October Supplement, selected years, 1972-2007.

ing to international benchmarks. The United States also has had the least sustained improvement in math and science from 1995 to 2007. It has, in fact, shown a 3-point decrease in the average science score for 4th grade science. The largest increase was in 8th grade mathematics, with an average score difference of 16 points.

The 2007 TIMSS report showed that African American and Hispanic students were narrowing the gap in 4th and 8th grade mathematics, but, as Figure 3-2 for Grade 8 shows, a large gap remained. Meanwhile, there is no consistent trend in science for either grade. In addition, as shown in Figure 3-3, at least at the 8th grade level, there is a large gap among schools by poverty level.

As shown in Figure 3-4, the Education Trust conducted an analysis of TIMSS data that shows that average mathematics and science scores for underrepresented minorities are below the national average and thus even less competitive globally. There is a larger gap between Hispanic/Latino and African Americans in mathematics and science for grades 4 and 8, except

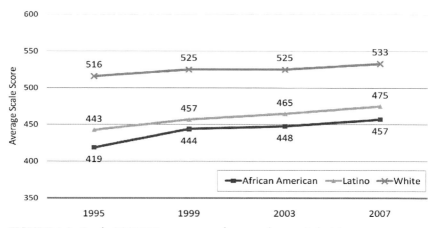

FIGURE 3-2 Grade 8 TIMSS average math scores by race/ethnicity.
SOURCE: The Education Trust. 2008. *Highlights from the Trends in International Mathematics and Science Study* (TIMSS) *2007*, Natonal Center for Education Statistics, U.S. Department of Education.

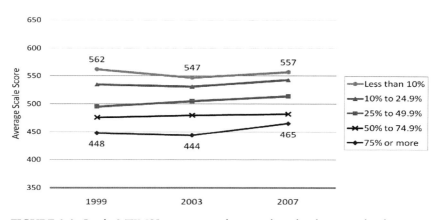

FIGURE 3-3 Grade 8 TIMSS average math scores by school poverty level.
SOURCE: The Education Trust. 2008. Highlights from Trends in International Mathematics and Science Study (TIMSS) 2007, National Center for Education Statistics, U.S. Department of Education.

in 4th grade science, where the average scores are about the same. African Americans scored lower than any group across the board.

The United States also compares its education system to that of the other Group of Eight (G-8) countries—Canada, France, Germany, Italy, Japan, the Russian Federation, the United Kingdom—that are among the world's

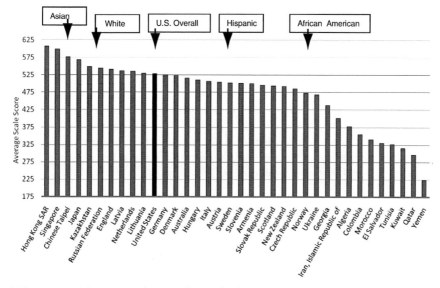

FIGURE 3-4 TIMSS Grade 4 math racial/ethnic subgroup comparison to all participating countries.
SOURCE: The Education Trust. 2008. *Highlights from Trends in International Mathematics and Science Study* (TIMSS) *2007*, National Center for Education Statistics, U.S. Department of Education.

most economically developed and among the nation's major competitors. *Comparative Indicators of Education in the United States and Other G-8 Countries: 2009*[4] shows that the United States has the largest percentage of 5- to 19-year-olds of all of the G-8 countries and experienced the highest growth in that subpopulation between 1996 and 2006. However, other G-8 countries outpace the United States in reading literacy, mathematics, and science. The United States also displays the widest disparity among racial/ethnic subgroups. The average years of teaching experience among 4th grade teachers in England and the United States was lower than in all other participating G-8 countries. The average teaching experience was three years lower in 2006 compared to 2001. While it spent a higher percentage of its GDP on education in 2005, it awarded among the lowest percentages of first university degrees in STEM of all the G-8 countries. It was the only G-8 country to award more first university degrees in the arts and humanities than in science, mathematics, and engineering.

---

[4] D. C. Miller, A. Sen, L. B. Malley, and S. D. Burns. 2009. *Comparative Indicators of Education in the United States and Other G-8 Countries: 2009* (NCES 2009-039). (Washington, DC: National Center for Education Statistics, Institute of Education Sciences, U.S. Department of Education.

Andreas Schleicher, in commenting on international benchmarking, indicated in a July 2009 briefing at the Woodrow Wilson International Center for Scholars that the U.S. education system needs a paradigm shift, one that embraces diversity, delivers equity, adopts universal high standards, and uses data and best practices. He commented that the distinction between public and private schools does not matter too much and that the United States should move from prescribed forms of teaching and assessment toward more personalized learning. All agree that the trends shown in the reports will only worsen if the nation does not aggressively and systematically remedy the problems that perpetuate the achievement gaps and underrepresentation of minorities in STEM.

Understanding Mathematics and Science Achievement Gaps

The achievement gap between white and minority students in K-12 mathematics and science is well documented in numerous research and statistical reports (e.g., *Condition of Education, The Nation's Report Card, Science and Engineering Indicators*). These confirm that family and community differences and school context have a significant impact on student achievement throughout the K-12 spectrum. For example, gaps in mathematics and science start in kindergarten and widen over time among underrepresented minorities generally, and especially among children with such risk factors as poverty, having a mother whose highest level of education was less than a high school diploma, or a home language other than English. *The Condition of Education 2009* reports that a higher percentage of white children had family members who read to them daily than did children of other racial/ethnic groups. Also, a higher percentage of Asian children were read to than Hispanic and American Indian/Alaska Native children at all ages, and than black children at ages two and four. Overall, a smaller percentage of children in poverty were read stories to, told stories to, or sung to daily by a family member, compared with children not in poverty.[5]

The achievement gap in mathematics and science is documented in numerous national assessments of student progress that have reported some fluctuations but the same trend for decades. As an illustration, Table 3-1 shows the average mathematics scores of students from the Early Childhood Longitudinal Survey (ECLS) by race/ethnicity from kindergarten to grade five for 1998, 2000, 2002, 2004, and 2007 as reported by the National Science Foundation (2008).[6] From kindergarten to 8th grade, white students posted a gain of 116 points; Hispanics, a gain of 113 points; and blacks,

---

[5] M. Planty, W. Hussar, et al. 2009. The Condition of Education 2009 (NCES 2009-081). National Center for Education Statistics Institute of Education Sciences, U.S. Department of Education, Washington, DC.

[6] National Science Foundation, *Science and Engineering Indicators 2008*.

TABLE 3-1 Average Mathematics Scores of Students from Beginning Kindergarten to Grade 8, by Race/Ethnicity: 1998, 2000, 2002, 2004, and 2007

Race/Ethnicity	Fall 1998 Kindergarten	Spring 2000 Grade 1	Spring 2002 Grade 3	Spring 2004 Grade 5	Spring 2008 Grade 8
All students	26	62	99	123	139
White	29	66	106	129	145
Black	22	52	84	105	123
Hispanic	22	56	92	118	135
Asian	30	65	105	133	148
Other[a]	25	59	95	120	137

[a]Includes non-Hispanic Native Hawaiians, Pacific Islanders, American Indians, Alaska Natives, and children of more than one race.
SOURCES: National Center for Education Statistics, Early Childhood Longitudinal Study, Kindergarten Class of 1998 and spring 2000, 2002, 2004, and 2007; and National Science Foundation, Division of Science Resources Statistics, special tabulations, *Science and Engineering Indicators 2010*.

a gain of 101 points. By 5th grade, the gap between white and black students in average mathematics scores was 24 points, and the average score of black 5th grade students was equivalent to the average 3rd grade score of white students.

ECLS data suggest that some gaps widened as students progressed through elementary school and that other gaps, such as those between boys and girls, emerged that were not present when students started school. Boys and girls started kindergarten at the same overall mathematics performance level, but by the end of 5th grade, boys had made larger mathematics gains than girls, resulting in a gender gap of four points.

Some research suggests that widening achievement gaps as students progress through school are, at least in part, a result of differential learning growth and loss during the summer (Cooper, 1996; Alexander et al., 2007). Most students lose about two months of grade level equivalency in mathematical computation skills over the summer months. Low-income students also lose more than two months in reading achievement, despite the fact that their middle-class peers make slight gains. These findings have been attributed to greater ability among higher-income parents to provide their children with mathematically stimulating materials and activities during the summer.

According to the Education Longitudinal Study (ELS), similar gaps persist through high school. For example, the proportion of 12th grade students overall demonstrating proficiency in advanced mathematics was lower and decreased as more advanced skills were tested. While each demographic subgroup examined improved in mathematics skills from 10th to 12th

TABLE 3-2 Average Science Score of Students in Grades 4, 8, and 12, by Race/Ethnicity: 1996, 2000, and 2005

Race/Ethnicity	1996	2000	2005
All Grade 4	147	147	151
White	158	159	162
Black	120	122	129
Hispanic	124	122	133
Asian/Pacific Islander	144	NA[a]	158
American Indian/Alaska Native	129	135	138
All Grade 8	149	149	149
White	159	161	160
Black	121	121	124
Hispanic	128	127	129
Asian/Pacific Islander	151	153	156
American Indian/Alaska Native	148	147	128
All Grade 12	150	146	147
White	159	153	156
Black	123	122	120
Hispanic	131	128	128
Asian/Pacific Islander	147	149	153
American Indian/Alaska Native	144	151	139

NOTES: Scores on 0-300 scale for each grade. In 2005, NAEP science assessment completed transition to an accommodations-permitted test.
[a]NA = not available. Special analyses raised concerns about accuracy and precision of national grade 4 Asian/Pacific Islander results in 2000; therefore omitted from National Center for Education Statistics (NCES).
SOURCES: NCES, *The Nation's Report Card: Science 2005* (NCES 2006-466) (2006); NAEP, 1996, 2000, and 2005 science assessments; and National Science Foundation, Division of Science Resources Statistics, special tabulations.

grade, minority students' scores were lower than those for white students (Table 3-2). By 12th grade, the average performance of black students was slightly lower than the average 10th grade performance of white and Asian students. A similar pattern is shown also for science assessments from 3rd through 12th grade. Thus, as larger numbers of underrepresented minorities are entering the STEM pipeline, many still are not progressing at a rate comparable to that of whites.

The National Assessment of Education Progress (NAEP) is the primary source used to report student performance data for the nation and specific geographic regions of the country and to produce *The Nation's Report Card*. The NAEP mathematics and science frameworks are developed under the direction of the National Assessment Governing Board, which sets specific achievement levels (basic, proficient, and advanced) for each subject area and grade as standards for student performance. The assessment uses two dimensions of mathematics, content areas and mathematical complexity. The science framework emphasizes assessing science concepts and application of scientific knowledge and skills rather than factual knowledge.

NAEP Mathematics

The most recent NAEP assessments of educational progress for 4th and 8th graders in mathematics show that all racial/ethnic groups showed higher average mathematics scores in 2009 than in 2007 and 1990.[7] Asian/Pacific Islander 4th grade scores were the highest followed by those of whites. Score increases did not consistently result in a significant closing of performance gaps between white and underrepresented minority students, although gains over the years for black students resulted in a smaller gap between black and white students in 2009 than in 1990. Male students continued to score two points higher on average than female students.

The average mathematics score for 4th graders in public schools (239) was lower than for students in private schools overall (246) and in Catholic schools specifically (245). Students who were eligible for free or reduced-price lunch continued to score lower on average than students who were not; however, average mathematics scores were higher in 2009 than in 2007 for all three groups. Mathematics scores increased from 2007 to 2009 for black students in Delaware and New Jersey; Hispanic students in Delaware, Florida, Missouri, and New Mexico; American Indian/Alaska Native students in Oklahoma. In no state did scores decline since 2005 for students overall or for any racial/ethnic group.

Table 3-3 compares the 2007 average scale scores and achievement level results by race/ethnicity for 4th and 8th grade public school students. Eighth graders reported gains for each of the five content areas. The largest percentage of the 168 questions that made up the 8th grade mathematics assessment (approximately 30 percent) focused on algebra. The percentages of 8th grade public school students at or above basic and proficient and advanced increased steadily from 1990 to 2007. White, black, and Hispanic students showed higher average mathematics scores in 2007 than in all previous assessment years. The score for Asian/Pacific Islander students showed no significant change in comparison to 2005 but was higher than in 1990. No significant change in the score was seen for American Indian/Alaska Native students.

---

[7] *The Nation's Report Card: Mathematics 2009. National Assessment of Educational Progress at Grades 4 and 8* (NCES 2010-451). 2010. Washington, DC: National Center for Education Statistics, U.S. Department of Education.

TABLE 3-3 Average Mathematics Scale Scores and Achievement Level Results by Race/Ethnicity for 4th and 8th Grade Public School Students, 2007

	Percentage of 4th Grade Students				
	Average Scale Score	Below Basic	At or Above Basic	At or Above Proficient	At Advanced
White	248	9	91	51	8
Black	222	37	63	15	1
Hispanic	227	31	69	22	1
Asian/Pacific Islander	254	9	91	59	16
American Indian/ Alaska Native	229	28	72	26	3

	Percentage of 8th Grade Students				
	Average Scale Score	Below Basic	At or Above Basic	At or Above Proficient	At Advanced
White	290	19	81	41	9
Black	259	53	47	11	1
Hispanic	264	46	54	15	2
Asian/Pacific Islander	296	18	82	49	17
American Indian/ Alaska Native	265	44	56	17	2

SOURCE: NCES. *The Nation's Report Card: Mathematics 2009* (NCES 2010-451), National Center for Education Statistics, U.S. Department of Education.

The most recent mathematics assessment for 12th graders is reported in *Science 2005* [8] and is based on a new framework. The assessment includes more questions on algebra, data analysis, and probability to reflect changes in high school mathematics standards and coursework. Sixty-one percent of high school seniors performed at or above the *Basic* level, and only 23 percent performed at or above *Proficient*.

Asian/Pacific Islander students scored higher than students from other racial/ethnic groups. The average for white students was 31 points higher than for black students and 24 points higher than for Hispanic students. Male students scored higher on average than female students overall. Thirty-six percent of Asian/Pacific Islander and 29 percent of white students scored at or above *Proficient*, while just 6 percent of black, 8 percent of Hispanic,

[8] W. Grigg, M. Lauko, and D. Brockway. 2006. The Nation's Report Card: Science 2005 (NCES 2006-466). U.S. Department of Education, National Center for Education Statistics. Washington, DC: U.S. Government Printing Office.

and 6 percent of American Indian/Alaska native students performed at that level. Fifty-five percent of students who reported taking a mathematics Advanced Placement course performed at that level.

The trends shown in long-term assessments in mathematics in 2007-2009 for students ages 9, 13, and 17 report similar results. In 2008, public school students scored lower than their private school counterparts at ages 9 and 13. Public school students scored lower than Catholic school students at all three ages in 2008.

From 2004 to 2008, black and Hispanic students ages 9 and 13 showed no significant change. At age 17, the score for neither white, black nor Hispanic students showed a significant change. Further, the gap between the white and the black and Hispanic students has narrowed since 1973 but has not changed significantly since 2004.

Significantly, taking higher-level mathematics courses was associated with higher scores on the long-term trend mathematics assessment in 2008 at ages 13 and 17 and in the main mathematics assessments for grades 4, 8, and 12.

NAEP Science

Table 3-4 compares the average science scale scores for each racial/ethnic group for 2000 and 2005. In 2005, the average 4th grade science score was higher than in previous assessment years, with underrepresented minorities and lower-income students making significant gains. Average science scores for 8th and 12th graders remained unchanged. From 2000 to 2005, black and Hispanic students' science scores improved, except for 12th graders, and the gaps between white and black and white and Hispanic students narrowed. However, there is still a 33-point score gap between white and black students, a 29-point score gap between white and Hispanic students, and a 27-point score gap between white and American Indian/Alaska Native students.

For both the 4th and 8th graders, a larger proportion of students eligible for free/reduced-price lunch were below the *Basic* level of proficiency than those who were not eligible. The 12th graders who scored at or above *Proficiency* tended to have at least one parent who graduated from college and 40 percent took at least one Advanced Placement science course. Within each course-taking level, male students outperformed female students. White and Asian/Pacific Islanders scored higher than black and Hispanic students.

Causes and Interventions

Researchers offer many explanations for the persistent achievement gaps while recognizing that there are many interrelated factors. They agree

TABLE 3-4 Average Science Scale Scores by Race/Ethnicity and Grade: 2000 and 2005

Grade and Year	Total[a]	White	Black	Hispanic	Asian/ Pacific Islander	American Indian/ Alaska Native
4th grade						
2000	147	159	122	122	[b]	135
2005	151	162	129	133	158	138
8th grade						
2000	149	161	121	127	153	147
2005	149	160	124	129	156	128
12th grade						
2000	146	153	122	128	149	151
2005	147	156	120	128	153	139

NOTE: Scale score ranges from 0 to 300. For a discussion of the science scale score definitions, please see http://nces.ed.gov/nationsreportcard/science/scale.asp. Race categories exclude persons of Hispanic ethnicity.
[a]Total includes race/ethnicity categories not separately shown.
[b]Reporting standards not met.
SOURCE: U.S. Department of Education, National Center for Education Statistics, National Assessment of Educational Progress (NAEP), 2000 and 2005 Science Assessments, retrieved January 30, 2008, from http://www.nces.ed.gov/nationsreportcard/nde.

that family and community differences, school context, low expectations and lack of exposure to role models, insufficient information about career opportunities, and availability of advanced courses affect minority students' success in mathematics and science. They attest also to the impact of interventions in promoting high achievement for minority students, notably The Algebra Project, Knowledge Is Power Program (KIPP), For Inspiration and Recognition of Science and Technology (FIRST) Program, Advancement Via Individual Determination (AVID) Program, and the Indigenous Education Institute. (See Boxes 3-2, 3-3, 3-4, and 3-5.)

Experts cite the need for additional research on effective interventions to eliminate racial achievement gaps as well as the need to facilitate the translation of research into practice. To address the latter problem, the superintendents from Evanston Township High School, Shaker Heights (Ohio), Chapel Hill (North Carolina), Arlington (Virginia), Ann Arbor (Michigan), Madison (Wisconsin), and nine other districts formed the Minority Student Achievement Network (MSAN) to collectively "create a body of educational research that informs classroom- and system-level practice and helps eliminate racial achievement gaps" and to "disseminate and implement effective

---

BOX 3-2
Knowledge Is Power (KIPP) Program

KIPP is a national network of free, open-enrollment, college preparatory public schools with a track record of preparing students in underserved communities for success in college and in life. There are currently 66 KIPP schools in 19 states and the District of Columbia serving nearly 21,000 students, 90 percent of whom are Hispanic or African American.

KIPP began in 1994 when two teachers, Mike Feinberg and Dave Levin, launched a 5th grade public school program in inner-city Houston, after completing their commitment to Teach for America. In 1995, Feinberg remained in Houston to lead KIPP Academy Middle School, and Levin returned home to New York City to establish KIPP Academy in the South Bronx. These two academies became the starting point for a growing network of schools that are transforming the lives of students in educationally underserved communities and are redefining the notion of what is possible in public education.

The majority of KIPP schools are middle schools, although the program is expanding to a Pre-K through 12 model. The KIPP middle school model has a proven track record of increasing student achievement, as measured by both national norm-referenced and state criterion-referenced exams. All KIPP schools share a core set of operating principles known as the Five Pillars: High Expectations, Choice & Commitment, More Time, Power to Lead, and Focus on Results. Eighty-five percent of the students matriculate to college.

SOURCE: http://www.kipp.org.

---

practices learned or developed by the MSAN to network members."[9] The group was formed in response to a National Research Council report that focused on the need for research that addresses the problems of educational practice.[10] The report proposed the establishment of a Strategic Education Research Program (SERP) that would tap the energies of researchers, practitioners, and policy makers to address fundamental issues in education, including how advances in research on human cognition, development, and learning can be incorporated into educational practice.

During the 2006-2007 school year, MSAN brought together teachers, social psychologists, and mathematics education researchers from the Charles A. Dana Center at the University of Texas to develop a compre-

---

[9] Laura Cooper. 2007. Why closing the research-practice gap is critical to closing student achievement gaps, *Theory and Practice* 46(4):317-324.

[10] National Research Council. 1999. *Improving Student Learning: A Strategic Plan for Education Research and Its Utilization.* Washington, DC: National Academy Press.

BOX 3-3
For Inspiration and Recognition of Science and Technology (FIRST) Program

FIRST is a nonprofit organization that engages K-12 students in mentor-based programs that develop STEM skills, motivate inquiry, and cultivate personal capabilities such as self-confidence, communication, and leadership. The programs combine interdisciplinary teamwork and competitions whereby students in four different age groups fund, design, build, and compete with robots in local, national, and international contests. The 2008-2009 programs are as follows:

- FIRST Robotics Competition for high school students
- FIRST Tech Challenge for high school students
- FIRST LEGO League for 9- to 14-year-olds
- Junior FIRST LEGO League for 6- to 9-year-olds
- FIRST Place for ages 6 to adult

Teams are diverse, including underrepresented minorities (56 percent), women (41 percent), students from families with limited educational background, and low-income populations. FIRST is supported by a network of over 3,000 corporations, educational and professional institutions, and individuals. The 2009 FIRST Robotics competition involved 42,000 high school students. The program awarded over $9.7 million in college scholarships.

SOURCE: http://.usfirst.org/aboutus/content.aspx?id=46.

---

hensive approach (the AYD Initiative) to introductory algebra for 9th grade students who previously have struggled with math. In addition to identifying the components of a strong curricular, instructional, and assessment design, the project focused on the social and psychological factors that affect student learning. The AYD Initiative (1) crosses traditional disciplinary boundaries, bringing together researchers with expertise in mathematics with researchers who study social and psychological factors—such as stereotype threat—that affect student achievement, (2) spans the research-practice gap, bringing together math and science teachers with social psychologists and mathematics researchers, and (3) utilizes a network of schools to disseminate successful instructional practices, arranging for teachers to observe each other's classes and to collaborate by sharing curricular materials and instructional strategies. Research finds that the most promising approaches to improving the low performance of certain groups of students pay as much attention to the social forces operating in schools and in classrooms as they

---

BOX 3-4
Advancement Via Individual Determination (AVID) Program

Advancement Via Individual Determination (AVID) is a program designed to help underachieving middle and high school students prepare for and succeed in postsecondary education. This program offers a rigorous program of instruction in academic "survival skills" and college level entry skills, such as how to study, read for content, take notes, and manage time. In addition to this, students participate in collaborative study groups or tutorials led by tutors who use skillful questioning to bring students to a higher level of understanding. Many of the AVID participants are underrepresented minorities and the first in their families to attend college, and many are from low-income or minority families.

Currently, AVID is offered in more than 3,500 schools in 45 states and 15 countries, including Department of Defense schools in Europe and the Pacific. This program has been found to work at a variety of schools in large urban areas, tiny rural towns, resource-rich suburban schools, and struggling communities. Total enrollment for AVID programs has reached about 300,000 students worldwide. Many AVID students take AP classes, completing their college eligibility requirements and getting into four-year colleges more often than students who don't take AVID. Of the high school participants, approximately 95 percent enroll in college, with more than 60 percent enrolled in four-year colleges and 86 percent rate of retention for all enrollees. AVID also helps ensure students, once accepted to college, possess the higher-level skills they need for college success.

To date, one of the most impressive and consistent indicators of AVID's success is the rate at which it sends students to four-year colleges. Seventy-eight percent of 2008 AVID graduates were accepted to a four-year college. Given this success, policy makers and school administrators now consider AVID an essential strategy for closing the achievement gap and making the college dream accessible to all students.

SOURCE: http://pac.dodea.edu/edservices/educationprograms/avid.htm.

---

do to skill and knowledge development. Research also provides evidence that social psychological interventions can have remarkably strong effects on engagement, as well as on test scores and grade point average.[11]

MSAN has developed a validation process to identify programs that are proven to be successful. The validation process is a peer review process that includes a review of multiple years of achievement and other quantitative data and an on-site visit by MSAN representatives to gather qualitative data.

---

[11] G. L. Cohen, J. Garcia, N. Apfel, and A. Master. 2006. Reducing the racial achievement gap: A social-psychological intervention. *Science* 313:1307-1310.

BOX 3-5
Indigenous Education Institute

The Indigenous Education Institute (IEI) was created in 1965 as a nonprofit 501(c)(3) institute with a mission to preserve, protect, and apply traditional indigenous knowledge in a contemporary setting, that of indigenous peoples today. IEI has developed numerous projects that preserve traditional knowledge, protect the knowledge in terms of indigenous protocol, and apply it to areas such as astronomy and other science disciplines.

IEI works closely with many indigenous organizations and institutions and with mainstream universities and K-12 schools. It develops educational materials such as the *Dine (Navaho) Universe;* a CD of Navaho astronomy *Stars Over Dine Bikeyah;* a cross-cultural astronomy book, *Sharing the Skies: Navaho Astronomy— A Cross Cultural View*, with comparisons of Navaho, Greek, and NASA Space Science worldviews; and *Guidebook to Navaho Astronomy* for the Starlab Portable Planetarium. IEI has developed a Dine Cosmic Model: "Strategic Planning and Evaluation in Accordance with the Natural Order" as perceived by the Navaho. IEI is known for development of place-based curriculum relevant to indigenous communities, such as Traditional Indigenous Geography, a traditional introduction to GIS technology.

The work of IEI is focused on the boundaries between traditional indigenous science and western science in formal and informal settings. The work of IEI is centered on the task of helping young native people find their own sense of self-identity and self-esteem in the world today, based on a firm foundation of thousands of years of cultural knowledge.

Utilizing effective practice in indigenous education, IEI researchers and educators engage diverse audiences with indigenous learning styles, using a holistic indigenous pedagogy in a variety of settings that include reservation schools, Native Hawaiian immersion schools, Native American educational leadership institutes, informal education settings such as museums and community centers, indigenous higher education institutions, mainstream scientists interested in exploring worldviews, and other indigenous and mainstream education and research institutions.

Examples of activities include: (1) K-12 science classes at Union Gap Elementary School, Union Gap, Washington, and Kula Kaipuni O Anuenue (Native Hawaiian Immersion School), Honolulu, Hawaii, and (2) workshops for educators, school boards, and administrators, including New Mexico science teachers, Utah Title 7 teachers, Navaho Nation science teachers, and NASA Explorer Schools.

SOURCE: http://www.indigenouseducation.org/about.html.

BOX 3-6
The El Paso Collaborative for Academic Excellence

The El Paso Collaborative for Academic Excellence, based at the University of Texas at El Paso, is a broad-based, citywide collaboration of education, business and civic leaders that has worked for over 17 years to transform schooling and ensure academic success for all young El Pasoans. From the beginning, the Collaborative's approach to reform has been grounded in the belief that all children, regardless of their race or ethnicity, the school they attend, or the neighborhood they live in, are entitled to a first-rate education, to educators who believe in them, and to a real chance to learn challenging content.

The Collaborative works with three urban and nine rural school districts and almost 200 schools supporting systemic education improvement and has focused on STEM teaching and learning, in particular. Its program priorities emphasize teacher and administrator professional development that is intensive, long term, and site based; rigorous and aligned curriculum, instruction, and assessment; building school and district organizational capacity to ensure high quality teaching and learning for all students; and development and implementation of policies that will ensure the work for the long term. It also works with the El Paso Community College and UTEP supporting transformed systems for preparing teachers and engagement of STEM disciplinary faculty in working with K-12 to improve STEM teaching and learning.

As a result of this work, achievement among all groups of students has increased greatly, the achievement gap between groups of students has declined significantly, and the high school graduation/college preparation levels exceed those of all other urban areas in the state.

- In 2007, the high school completion rate of students in El Paso's three urban districts was 77 percent, the highest among all major Texas cities, including Austin, Dallas, and Houston.
- In 1993, just 32 percent of African American and 36 percent of Hispanic students achieved passing scores on the math portion of TAAS, the Texas statewide assessment. By 2008, 77 percent of all students passed the much more demanding TAKS (Texas Assessment of Knowledge and Skills), and the achievement gap was reduced significantly among all groups of students.
- Enrollments and pass rates in college preparatory courses have risen dramatically. Key STEM courses provide good examples. In 1993, just 63 percent of students were enrolled in Algebra I, with 59 percent passing. By 2007, 100 percent of students were enrolled in Algebra I, with 74 percent passing the course. In the critically important course, Algebra II, enrollments jumped from 42 percent in 1993 to 98 percent in 2007, with a concomitant increase in pass rates. Enrollments in Chemistry tripled from 28 percent in 1993 to 84 percent in 2007, with pass rates also increasing.

The Collaborative has become a national model of citywide efforts to recreate schools; has been featured in numerous national publications, including *Education Week* and *School Leader*; and has been awarded over $60 million in grants from the National Science Foundation, the U.S. Department of Education, the Pew Charitable Trusts, and others, for its work to bring about K-16 systemic education reform.

SOURCE: Diana Natalicio, President, University of Texas at El Paso.

The El Paso Collaborative for Academic Excellence (Box 3-6) is another example of a multifaceted intervention for school reform with proven success. Based at the University of Texas at El Paso, this initiative has become a national model.

Informal Science Education: Seamless Networks

Increasingly, informal science education is being used to address issues of scientific literacy, cultural relevance, equity, and access for women and minority populations. The National Science Foundation was the first to recognize and support the role of community organizations, museums, and media as rich resources and essential partners in the educational process. It created the Division of Informal Science Education (ISE) in 1984 based on the recommendation of *Educating Americans for the 21st Century: A Report to the American People and the National Science Board.*[12]

"An important perspective on informal science learning in informal environments emphasizes that, although treating the construct of culture as a homogeneous categorical variable is problematic, people nonetheless do 'live culturally'."[13] Informal science education can make science accessible, meaningful, and relevant for diverse students by connecting their home and community cultures to science. Nancy Brayboy and David Begay (2005) demonstrate in *Sharing the Skies* how to bridge the culture of science through a cross-cultural view of Navajo astronomy.

More research is needed on how to structure informal science education to better serve minorities; however, designed environments such as museums can provide access to specific content and facilitate social interaction and learning in intergenerational groups. Research has documented that participation in many venues (e.g., designed formal settings, science media) is skewed toward the dominant cultural group, although there are some exceptions.

> Wheaton and Ash's research (2008) on science education in informal programming with Spanish-speaking families found that participating girls welcomed and enjoyed the bilingual program because they learned science terminology and concepts in both languages and thus better communicated with their parents (who were predominantly Spanish speaking) about what they were doing and learning in camp.[14]

---

[12] ISE is now a program unit in the Division of Research on Learning in Formal and Informal Settings.

[13] National Academies. 2009. *Learning Science in Informal Environments: People, Places, and Pursuits.* Washington, DC: The National Academies Press, p. 210.

[14] National Academies. *Learning Science*, p. 234.

The Native Waters and the Algebra Projects are cited as examples of approaches that incorporate a learner's cultural identity.

In sum, an informal environment designed to serve particular cultural groups and communities should be developed and implemented with the interests and concerns of these groups in mind. Project goals should be mutually determined by educators and the communities they serve.[15]

THE K-12 SPECTRUM

K-12 is considered a continuum, and the problems at one level affect each succeeding level. Moreover, the issues pertaining to the achievement and progression of underrepresented minorities are common to all levels. The major issues are described below.

Preschool

Pre-kindergarten (pre-K) is designed to prepare children for entry into elementary school by cultivating the prerequisite developmental skills for success in the early grades, and the long-term benefits of high-quality early childhood programs for all children are well documented. Studies also cite the positive effects on absenteeism, classroom behavior, grade repetition, high school graduation rates, crime, and academic achievement, substantially countering the negative effects of family and environmental risk factors for low-income and minority populations. There is evidence also that the benefits of investments in pre-K outweigh the costs to society (Bartik 2006, Dickens et al. 2006). However, in the United States there is a fragmented approach to early childhood programs and services, and children who have the most risk factors still do not enter kindergarten with the intellectual and social tools they need to progress successfully through elementary school.

Head Start is the nation's primary program for addressing the educational and developmental needs of children of low-income families who do not otherwise have access to quality preschool education. It is a "national program that promotes school readiness by enhancing the social and cognitive development of children through the provision of educational, health, nutritional, social and other services to enrolled children and families."[16] In 2007, the program served over 900,000 children at a total cost of about $6.888 billion, or about $7,326 per child. However, because of chronic underfunding and recent budget cuts, it enrolls only about 40 percent

---

[15] Ibid, p. 235.

[16] http://www.acf.hhs.gov/programs/ohs/about/index.html#factsheet (accessed March 4, 2009).

And we should raise the bar when it comes to early learning programs. . . . Today, some early learning programs are excellent. Some are mediocre. And some are wasting what studies show are—by far—a child's most formative years.

That's why I have issued a challenge to America's governors: If you match the success of states like Pennsylvania and develop an effective model for early learning; if you focus reform on standards and results in early learning programs; if you demonstrate how you will prepare the lowest income children to meet the highest standards of success—then you can compete for an Early Learning Challenge Grant that will help prepare all our children to enter kindergarten ready to learn.

—President Barack Obama
Remarks to the NAACP, July 16, 2009

of eligible children.[17] With the projected increase in minority pre-school population, this means that fewer children will have access to the program. The recently enacted Stimulus Act[18] provides a one-time infusion of $1.1 billion to double the number of children served by Early Head Start over two years, an additional $1 billion to expand and improve Head Start, and an additional $2 billion in funding for the Child Care and Development Block Grant.[19] However, this level of funding will need to be sustained into the future.

Program statistics for FY 2007 show that 30.1 percent of enrolled children identify as Black/African American, 34.7 percent as Hispanic/Latino, 4.0 percent as American Indian/Alaska Native, and 0.8 percent as Pacific Islander. About 8.0 percent of program funding is targeted to American Indian-Alaska Native and Migrant and Seasonal Programs.

Many states also now provide pre-kindergarten programs as a result of the growing need for early intervention and documented evidence of its effectiveness. However, there is wide variation in these programs and great disparities exist from state to state and among districts, even in their stages of development. Some programs cover 3- and 4-year olds (usually 4-year-olds), while others target special populations such as the urban and rural poor. Some states offer full-day while others offer part-day programs. Some offer full-year programming, but most are part-year corresponding to the academic year. Some provide comprehensive developmental services to children and families; others focus more directly on just the academic

---

[17] http://www.results.org/website/download.asp?id=3801 (accessed March 4, 2009).

[18] The American Recovery and Reinvestment Act of 2009.

[19] http://www.whitehouse/gov/omb/assets/fy2010_new_era/Department_of_Health_and_Human_Service1.pdf.

preparation for kindergarten. Oklahoma is the only state where practically every child can start school at age four. Enrollment in state-funded pre-kindergarten increased to more than 1.1 million children in 2007-2008, with more than 973,178 4-year-olds alone. Thirty-three of the 38 states offering such programs increased enrollment.[20]

Total state spending for pre-K rose to almost $4.6 billion in 2007-2008. In most states, however, funding per child from all sources (local, state, and federal) appears too low for programs to meet the ten benchmarks for quality standards established by the National Institute for Early Education Research.

Advocates such as Pre-K Now are challenging state pre-K programs to provide universal comprehensive services, particularly to rural and underrepresented minority children. However, states are faced with fiscal constraints that limit the expansion, enhancement, and quality of pre-K programs. Some are creatively leveraging federal and state funds in order to offer Head Start and childcare providers and to support staff development. Some states have had success at funding pre-K as a line item in state budgets or as an element of their school formulas, but such efforts still rely heavily on federal dollars to supplement and intensify services.

Research and assessment reports document the efficacy of Head Start and state pre-k programs as well as the differences in outcomes between the two[21] (U.S. Department of Health and Human Services Administration for Children and Families January 2010 Head Start Impact Study; Gormley, Phillips, and Gayer, 2008; National Institute for Early Education Research; Barnett, Jung, Wong, Cook, and Lamy, 2007). They also cite a number of issues that impact the effectiveness of these strategies and argue for more collaboration and integration to optimize the services to children and families.

Teacher quality is a major issue for public pre-K programs and Head Start. In a multistate study of pre-K, the National Center for Early Development & Learning at the University of North Carolina at Chapel Hill found that about 81 percent of public pre-K teachers had a bachelor's degree or higher, and only 8 percent reported no postsecondary degree.[22] This compares to 57 percent with a bachelor's or higher in nonpublic school

---

[20] States offering no programs are Alaska, Hawaii, Idaho, Indiana, Mississippi, Montana, New Hampshire, North Dakota, Rhode Island, South Dakota, Utah, and Wyoming.

[21] The 2010 Head Start Impact Study found that by the end of the 1st grade, Head Start children did significantly better on the vocabulary measure and test of oral comprehension than non-Head Start children. The Abbott Preschool Program Longitudinal Effects Study (2009) showed that children who attended the Abbott pre-K continued to outperform their peers at the end of the second grade and that there are advantages for those who had two years of preschool compared to just one.

[22] Pre-K Education in the States. *Early Developments*, FPG Child Development Institute: University of North Carolina at Chapel Hill. (Spring 2005) 9(1):6.

settings and 24 percent with no postsecondary degree. African American and Hispanic/Latino teachers were somewhat less likely to have a bachelor's degree than white teachers. In addition, classrooms where the teacher did not have a bachelor's degree tended to have a higher proportion of children from low-income backgrounds than classrooms where the teacher had a bachelor's degree. The average salary received by teachers in this population (about $19 per hour) is higher than has been reported in studies of childcare teachers (somewhat over $8 per hour) or Head Start teachers (about $16 per hour). This is likely due to the higher education levels of these pre-K teachers compared to childcare or Head Start teachers.

Scholars point to the need to rethink the pre-K system for underrepresented minorities, using research and best practices concerning 3- and 4-year-olds and their families. As one concept, the FPG Child Development Institute at the University of North Carolina at Chapel Hill has launched a new model for First School with the following features:[23]

- Be available for all children from age three to about age eight.
- Provide seamless transitions for children from pre-K to 3rd grade.
- Integrate and align curriculum across grades.
- Provide developmentally appropriate facilities and activities.
- Focus on academic skills, social-emotional development, and physical health.
- Involve strong and meaningful partnerships with families in developing, implementing, and evaluating the model.
- Use data to drive and monitor school change.

The major issues confronting the nation in developing model preschool programs for underrepresented minority children are the following:

- **Funding:** Per-pupil funding is too low for many states to improve the overall quality of programs, and there is growing disparity in funding between states. The bulk of federal funding for early childhood education now goes to Head Start and to the Child Care Block Grant, which provides childcare subsidies for poor families. As these programs are not designed to serve all young children, a new federal initiative is needed to support early learning and development more broadly.
- **Teacher preparation and quality:** Teacher qualifications for many public pre-K and Head Start programs are inconsistent with those of K-12 schools, and their salaries are not comparable to the salaries earned by kindergarten teachers. Resources are inadequate to support pre-service teacher

---

[23] NCEDL Director's Notes, *Early Developments*, FPG Child Development Institute: UNC-Chapel Hill (Spring 2005) 9(1):5.

education and in-service professional development focused on the needs of minority populations. There are inadequate incentives to recruit and retain qualified pre-K teachers.

• **Access**: Head Start targets primarily children from low-income families. While the amount of state support for pre-K has increased overall, discrepancies persist between state programs and among individual programs within states. Leading states, such as Oklahoma, have enrolled more than 70 percent of the state's 4- year-olds in state-funded programs, while others serve fewer than 5 percent. Some states have yet to establish any publicly funded pre-K programs. Public pre-K programs and Head Start must respond to a growing diverse population of children, in terms of race and ethnicity, socioeconomic status, and dominant language.

• **Curriculum and standards**: Head Start is guided by federal guidelines. State standards vary, although most are benchmarking the National Institute for Early Education Research quality standards. There is a need for better connection to full-day kindergarten and primary grades with aligned standards and curricula in a coherent education program for pre-K through 3rd grade. Head Start and other pre-K programs have produced improvement in language and literacy skills but need to focus on early math, namely, solving simple word problems involving counting, simple arithmetic, and basic measurements. While well-designed pre-K does improve children's social and cognitive skills, gains for minority children diminish as they advance beyond kindergarten.

• **Assessment and data driven policies**: Research should drive practice, particularly as a means of directly addressing the achievement gap. Translating research to practice and replicating best practices are critical strategies.

Mathematics and Science Teacher Quality

Rising Above the Gathering Storm recommends aggressive actions to recruit and strengthen the training of mathematics and science teachers and to foster high-quality teaching with "world-class curricula, standards, and assessments of student learning." The report cites exemplars in these efforts—UTeach at the University of Texas and California Teach at the University of California, the Merck Institute for Science Education, University of Pennsylvania Science Teachers Institute, Advanced Placement Incentive Program, and Laying the Foundation. The report recommends also statewide specialty high schools and inquiry-based learning as means to increase the number of students who pass AP and IB science and mathematics courses in an effort to enlarge the pipeline of students who are prepared to enter college and graduate with a degree in science, engineering, or mathematics.

The No Child Left Behind Act also focuses on improving teacher quality and authorizes the Department of Education Academic Improvement and Teacher Quality Program. It requires "highly qualified"[24] teachers in all core academic classes and asks district and state leaders to attest that low-income and minority students are not taught disproportionately by out-of-field teachers. However, there is some discrepancy between the Consolidated State Performance Report and data from the Department of Education Schools and Staffing Survey suggesting that out-of-field teaching may be more prevalent than state reports indicate. It is clear that much more is needed to improve student achievement.

Public school teachers have been predominantly white. In 2008, African American and Hispanic teachers each represented 7 percent, and other racial/ethnic minority groups represented less than 2 percent.[25] The racial and ethnic distributions among middle school and high school mathematics and science teachers resemble that same pattern. This is a salient issue, because declines in the number of minority teachers affect both minority and majority children. "A quality education requires that all students be exposed to the variety of cultural perspectives that represent the nation at large."[26]

While the number of mathematics and science teachers has steadily increased since 1999, particularly in middle schools or in schools with the highest concentration of minority and poor students, students in non-minority and wealthier schools have continued to be substantially advantaged by the distribution of the teacher pool. In 2003, mathematics and science teachers with a master's degree or higher were more prevalent in low-minority schools than in high-minority schools. Fully certified teachers were also more common in schools with lower proportions of underrepresented minority and poor students. In addition, although the overall percentage of beginning teachers with practice teaching experience has dropped from 1999, fewer beginning mathematics and science teachers who had any practice teaching were in schools with large concentrations of minority and poor students. Beginning mathematics and science teachers who participated in practice teaching were more likely than their counterparts without any practice teaching to report feeling well prepared to perform various teaching tasks.

---

[24] To be considered "highly qualified," a teacher must possess a bachelor's degree and full state certification or licensure and demonstrate knowledge of the content in the subject he or she teaches.

[25] NCES. 2009a. *Characteristics of Public, Private, and Bureau of Indian Education Elementary and Secondary School Teachers in the United States From the 2007-08 Schools Staffing Survey* (NCES 2009-324). Washington, DC: National Center for Education Statistics, U.S. Department of Education.

[26] M. Donnelly. 1988. *Training and Recruiting Minority Teachers*, ERIC Digest Series Number EA29, p. 1.

Out-of-field teaching has received much attention.[27] Again, mathematics and science teachers in schools with higher concentrations of minority and poor students are more likely to be teaching out of field. In fact, in high-poverty schools, more than one in every four core classes (27.1 percent) has an out-of-field teacher, compared with only about half as many classes (13.9 percent) in low-poverty schools. These are the very schools where students desperately need good teachers. High-minority and high-poverty schools are also more likely to have new mathematics and science teachers. This is particularly true in middle schools. All indicators examined showed a general pattern of unequal access to the most qualified teachers: Low-minority and low-poverty schools were more likely than high-minority and high-poverty schools to have teachers with more education, better preparation and qualifications in their field, and more experience.

Significant work remains to be done to eliminate out-of-field teaching and guarantee that low-income and underrepresented minority students have teachers with demonstrated knowledge in their subject areas. Researchers report that out-of-field teaching does not necessarily result from teacher shortages or inadequate preparation of teachers but from poor planning or administrative convenience.[28] States should offer incentives to recruit and retain teachers into high-needs schools and not allow school districts to assign a greater number of out-of-field or new teachers to high-needs schools than the district average.

The NSF Robert Noyce Teacher Scholarship Program addresses this issue by providing funding to institutions of higher education for scholarships, stipends, and programmatic support to recruit and prepare STEM majors and professionals to become K-12 mathematics and science teachers. Scholarship and stipend recipients are required to teach for two years in a high-need school district[29] for each year of support. In addition, the program supports the recruitment and development of NSF Teaching Fellows who receive salary supplements while fulfilling a four-year teaching requirement and the development of NSF Master Teaching Fellows by providing professional development and salary supplements while they are teaching for five years in a high-need school district.

Teacher quality is considered the most critical factor affecting academic achievement.[30] Research by Harris and Sass (2008) and Ingersoll (2008) attests to the impact of teacher training and teacher quality on stu-

---

[27] "Out-of-field teachers" are defined as those possessing neither certification in the subject they have been assigned to teach nor an academic major in that subject.

[28] R. M. Ingersoll. 2008. *Core Problems: Out-of-Field Teaching Persists in Key Academic Courses and High-Poverty Schools*. Washington, DC: The Education Trust.

[29] Defined in Section 201 of the Higher Education Act of 1965 (20 U.S.C. 1021).

[30] There is no consensus on what defines teacher quality. The most common measures are content knowledge, experience, pedagogical skills, and academic skills and knowledge.

dent achievement in mathematics and science, particularly content-focused teacher professional development. They found that since experience greatly enhances the productivity of elementary and middle school teachers early in their careers, policies should be designed to promote retention of young teachers. In addition, advanced degrees are not correlated with the productivity of elementary school teachers; thus, current salary schedules, which are based in part on educational attainment, may not be an efficient way to compensate teachers. In addition, more resources should be directed toward content-focused training for teachers in the upper grades, and changes are warranted in professional development at the elementary level and in pedagogical in-service training generally. They found no evidence that education majors are significantly more productive as teachers than nonmajors, so it seems worthwhile to experiment with "alternative certification" programs that facilitate the entry into teaching of people with majors other than education. These researchers also suggest that more experienced teachers appear more effective in teaching elementary math and reading and middle school math.

The Science and Mathematics Teacher Imperative (SMTI) was formed as an ambitious effort by members of the Association of Public and Land Grant Universities to substantially increase the number and diversity of high quality mathematics and science teachers in middle schools and high schools. Through partnerships among universities, school systems, the business community, and state and federal governments, SMTI intends to respond to statewide needs for teachers on a sustained basis. SMTI is developing an analytic framework to capture and share leading evidence-based practices systematically with other institutions to enhance the quality of teachers. The National Math and Science Initiative also recommends keeping content knowledge the priority for elementary and secondary teachers and offers a guide for state policy makers to inventory their own policies and regulations to ensure that each contributes to solving the teacher pipeline problem.[31]

The Education Trust presents a plan for equity with immediate and longer-term steps to remedy the unfair distribution of teacher quality. The Education Trust presents a case study of how three states—Ohio, Illinois, and Wisconsin—and their three biggest school systems—Cleveland, Chicago, and Milwaukee—attempted to solve this problem.[32] The result of their surveys showed that the current system of distributing teacher quality produces exactly the opposite of what is needed to close achievement gaps. They found consistently that highly qualified teachers were more

---

[31] *Tackling the STEM Crisis: Five Steps Your State Can Take to Improve the Quality and Quantity of Its K-12 Math and Science Teachers.* National Math and Science Initiative. Available at http://www.nctq.org/p/docs/nctq_nmsi_stem_initiative.pdf.

[32] H. Peske and K. Haycock. 2006. *Teaching Inequality: How Poor and Minority Students Are Shortchanged on Teacher Quality.* Washington, DC: The Education Trust.

likely teaching in schools with less poverty and fewer students of color and in schools with higher achievement. Researchers at the Illinois Education Research Council looked at a combination of measures and documented differences in the combined characteristics of teachers in high- and low-poverty schools and attempted to understand how, if at all, these differences affected student achievement. They found that quality matters a lot. For example, in schools with just average teacher quality, students who completed Algebra II were more prepared for college than their peers in schools with the lowest teacher quality who had completed calculus.[33]

The federal government could use policy as a lever to address the equity problem. Title I—Improving the Academic Achievement of the Disadvantaged is "to ensure that all children have a fair, equal, and significant opportunity to obtain a high-quality education and reach, at a minimum, proficiency on challenging state academic achievement standards and state academic assessments." The assumption seems to be that these funds are added to an equitable base of state and local resources. However, the schools that have had the most low-income children have had the least qualified teachers who were paid less than veteran and fully credentialed teachers.[34] Thus, school districts could spend less money in Title I schools than in other schools even with the addition of Title I funds. The law requires "comparability" in the educational opportunities provided in Title I and non-Title I schools but ignores disparities in teacher qualifications across schools and the resulting disparities in teacher salaries. Thus, millions of dollars were directed away from high-poverty schools to subsidize higher teacher salaries in schools with fewer children from low-income families. Principals in high-poverty schools received no additional money to train and support their inexperienced, lower-paid staff. The comparability loophole allowed districts to not confront the discriminatory effects of the current system.

With the No Child Left Behind Act (NCLB), Congress insisted that states and districts had to commit to identifying and addressing shortages of qualified teachers in high-poverty and high-minority schools as a condition of continuing to receive federal funds to help with the education of disadvantaged students. Every state and district that wanted to participate in Title I had to develop a plan "to ensure that poor and minority students are not taught at higher rates than other children by inexperienced, unqualified, or out-of-field teachers."

U.S. Education Secretary Arne Duncan has argued for differential pay for teachers of mathematics, science, and other high-need subjects, stating

[33] Karen J. DeAngelis, Jennifer B. Presley, and Bradford R. White. 2005. Illinois Education Research Council. Policy Report: IERC 2005-1.

[34] Lindsey Luebchow. 2009. *Equitable Resources in Low Income Schools: Teacher Equity and the Federal Title I Comparability Requirement.* Washington, DC: Education Policy Program, New America Foundation.

> I believe that education is the civil rights issue of our generation. And if you care about promoting opportunity and reducing inequality, the classroom is the place to start. Great teaching is about so much more than education; it is a daily fight for social justice.
>
> —Arne Duncan, Secretary of Education,
> at the University of Virginia, October 9, 2009

that there needs to be a more market-driven approach to teacher pay in which schools can bid for outside talent and recruit it. "It's not the solution," he has said of this approach to addressing mathematics and science teacher shortages, "but it's a piece of the solution." He maintains further that teacher colleges need to become more rigorous and clinical, much like other graduate programs, in order create that "new army of great teachers." High-quality alternative pathways for aspiring teachers that should expand in coming years, Duncan contends, include those like the New Teacher Project, the Troops to Teacher Program, and Teach for America.

College Readiness

Each year high school students take the ACT and/or SAT in order to qualify for admission to college. However, the tests also provide compelling feedback about the academic preparation of students throughout the K-12 continuum. The 2009 SAT and ACT reports document and reaffirm the achievement gap between white and underrepresented minority students. Although there is considerable controversy about the validity of using either test to predict college success and the racial bias implicit in test design, the tests still are used as the standard to guide most college admissions decisions.

More than 1.5 million students in the class of 2009 took the SAT. Forty percent were underrepresented minority students, an increase from 38.0 percent in 2008 and 29.2 percent in 1999 and the largest and most diverse group ever to take the test. There was an increase also in the number of students who said they were first-generation college students and in the number who reported that English was not their first language.

Not surprisingly, average SAT scores vary widely by race, gender, and income, and some gaps even widened. In 2009, the average scores were 501 in critical reading, 515 in mathematics (same as 2008), and 493 in writing. The reading and writing scores each dropped by one point for all groups.

TABLE 3-5 Average Scores on the SAT Reasoning Test by Race/Ethnicity, 2009

Race/Ethnicity	Number	Percent	Reading Mean	Math Mean	Writing Mean
American Indian/Alaska Native	8,974	1	486	493	469
Asian/Pacific Islander	158,757	10	516	587	520
African American	187,136	12	429	426	421
Hispanic	206,584	14	453	458	446
White	851,014	56	528	536	517
Other	51,215	3	494	514	493
No Response	66,448	4	472	501	469
Total	1,530,128	100	501	515	493

NOTE: Separate scores for Mexican or Mexican American, Puerto Rican, and Other Hispanic, Latino, or Latin American are averaged in the row labeled Hispanic.
SOURCE: Total Group Profile Report, College Board, 2009 College-Bound Seniors.

The differences in SAT scores were most pronounced between Asian students, who scored an average of 1623 out of 2400, and African American students, who averaged 1276. The national average was 1509. Meanwhile, African American students had the lowest average combined mathematics and critical reading score of 855, while white students had an average combined score of 1064 (Table 3-5). Moreover, students with a reported family income of more than $200,000 increased their average combined score over 2008 by 26 points, to 1702. Students who reported family incomes of less than $20,000 a year averaged 1321, a gain of one point.

Females comprised 53.5 percent of the 2009 test-taking group and had a combined mathematics and critical reading score of 997 compared to 1037 for males. African American females and males had the lowest average combined mathematics and critical reading scores of 851 and 861, respectively, while Asian female and male students had the highest average combined scores of 1087 and 1118, respectively. White females and males ranked second, with combined scores of 1046 and 1085.

Mean scores in mathematics for underrepresented minorities vary considerably among the states, as shown in the sample in Table 3-6. States acknowledge the performance gaps for underrepresented minorities, and some have implemented interventions to improve mathematics achievement. For example, Georgia introduced a new mathematics curriculum, the Georgia Performance Standards (GPS), beginning with 6th graders in 2005. The GPS has been phased in one grade per year. Students in the class of 2012 will be the first graduating class to have been fully instructed in GPS mathematics during secondary school.

TABLE 3-6 Average State Mathematics Scores on the SAT Reasoning Test by Race/Ethnicity, 2009

Race/Ethnicity	Nation	CA	GA	MA	MI	NY	TX	WI
American Indian/Alaska Native	493	501	492	496	598	473	513	562
Asian/Pacific Islander	587	568	572	593	673	571	582	666
African American	426	428	422	430	484	419	436	510
Hispanic	458	458	480	457	546	439	473	548
White	536	549	522	539	604	536	543	612
Other	514	524	487	504	602	487	511	592
No Response	501	513	484	503	591	464	481	584

NOTE: Separate scores for Mexican or Mexican American, Puerto Rican, and Other Hispanic, Latino, or Latin American are averaged in the row labeled Hispanic.
SOURCE: Total Group Profile Report, College Board, 2009 College-Bound Seniors.

A record number of students took the ACT in 2009. Of the 1,480,000 who took the test, only about 23 percent were underrepresented minorities, and 64 percent were white. Overall test scores remained the same between 2005 and 2009, although 25 percent more high school graduates have taken the ACT over this period, and the group has become more heterogeneous. Average composite scores for all groups increased between 2005 and 2009 except for African American graduates, whose average score decreased by 0.1 scale point. The ACT establishes college readiness benchmarks and reported that students from most racial/ethnic groups met the English benchmark, followed in order by the reading, mathematics, and science benchmarks.[35] Three benchmarks were met by at least 50 percent of Asian American/Pacific Islander and white students, while one was met by at least 50 percent of American Indian/Alaska Native students. None of the benchmarks were met by at least 50 percent of Hispanic or African American students. As with the SAT, graduates who took a college preparatory core curriculum in high school were more likely to meet the ACT benchmarks in 2009. The largest curriculum-based difference in benchmark attainment rates was in mathematics.

Figure 3-5 compares the percentage of students taking core courses, by race/ethnicity, in 1999 and 2009. There is an increase from 74 to 80 percent of students overall completing core courses since 1999 with Native American (66 to 75) and white students (76 to 84) showing the largest gain. Black and Mexican American students are the least represented in 2009, with percentages of 72 and 71, respectively.

---

[35] ACT defines college readiness as students having approximately a 75 percent chance of earning a grade of C or higher in first-year college English composition; college algebra; history, psychology, sociology, political science, or economics; and biology.

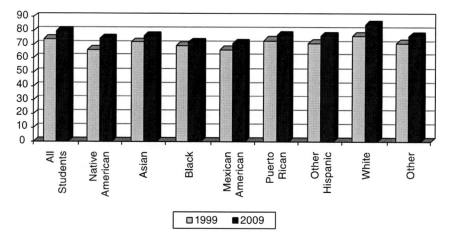

FIGURE 3-5 Percentage of students with core course work during high school by race/ethnicity.
SOURCE: The College Board, Graph Set 5: Course Taking Patterns Continued: 1999 and 2009.

The strongest SAT and ACT performers had three things in common. They had completed a core curriculum, had taken the most rigorous courses, and had familiarized themselves with the test. The core curriculum consisted of four or more years of English, three or more years of natural science, and three or more years of social science and history. Students in the SAT class of 2009 who took core curricula scored an average of 46 points higher on the critical reading section, 44 points higher on the mathematics section, and 45 points higher on the writing section than those who did not. Similarly, students in the class of 2009 who had taken the most demanding honors or Advanced Placement® courses had higher SAT scores on this year's test. For example, students who took AP® or honors English courses scored 60 points higher in critical reading and 59 points higher in writing than the average of all students. Similarly, students who took AP or honors math courses had a 79-point advantage compared to the average mathematics score. And students who had previously taken the Preliminary SAT/National Merit Scholarship Qualifying Test scored 121 points higher on average than those who did not take the test. The overall performance

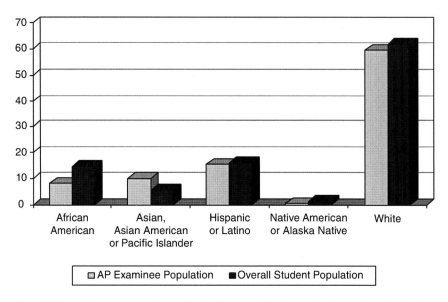

FIGURE 3-6 Access to AP by race/ethnicity—U.S. Public schools: High school class of 2009.
SOURCE: The College Board. 2010. *The 6th Annual AP Report to the Nation.*

of underrepresented minorities on the 2009 SAT and ACT is indicative of the trend seen for decades.

In a recent report, the College Board presented data showing that although there are increasing numbers of African American, Hispanic, and American Indian students participating in AP, these students still remain underserved and are less successful on AP exams, especially African Americans.[36] As shown in Figure 3-6, African American students represent 14.5 percent of the public school graduating class of 2009, and 8.2 percent of the AP examinees (compared to 14.4 percent and 7.8 percent in 2008). Hispanic students represent 15.9 percent of the public school graduating class of 2009 and 15.5 percent of the AP examinees (compared to 15.4 percent and 14.8 percent in 2008). Generally, states have done poorly in closing the equity and excellence gap for minority students, particularly the states with the largest percentage of underrepresented minorities in the 2009 graduating class. This further affirms that these students are not being ade-

---

[36] The College Board. 2010. *The 6th Annual AP Report to the Nation.* New York, NY: The College Board. The College Board uses an AP Exam score of three or higher to define success. More research is needed to establish the conditions under which AP Exam scores lower than three relate to college success.

quately prepared for success in college proportionately to white and Asian students. "Major initiatives are needed to ensure adequate preparation of students in middle school and 9th and 10th grades so that all students will have an equitable chance at success when they go on to take AP courses and exams later in high school."

The report cites the National Governor's Association's Advanced Placement Expansion Project and the National Math and Science Initiative's Training and Incentive Program as two major initiatives that are helping schools make progress toward closing the achievement gaps. They demonstrate the importance of state-level policies in expanding access to AP to more diverse students. For example, states with large Hispanic student populations, such as Florida, Texas, and California, all have AP-related multiyear student reform initiatives that use AP as a capstone. States with large African American student populations are only beginning to address these disparities. There is general consensus that the factors that contribute to better performance also impact college enrollment and completion. As there has been greater pressure for improved academic achievement from employers and colleges and universities, some states have increased the number of required mathematics and science courses, and all have adopted content standards in mathematics and science. However, there still is no alignment of high school graduation requirements and first-year college course requirements.

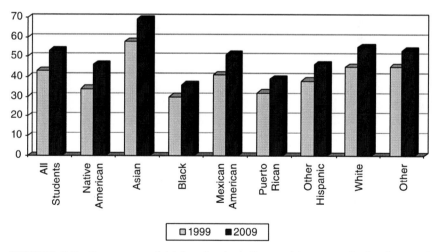

FIGURE 3-7 Percentage of high school students taking pre-calculus by race/ethnicity: 1999 and 2009.
SOURCE: The College Board, Graph Set 5: Course-Taking Patterns: 1999 and 2009.

Generally, more high school students have completed more mathematics and science courses since 1990, including more advanced courses. However, an increase in course taking is not sufficient to significantly increase the overall performance of underrepresented minorities.

Student success in mathematics is among the most reliable predictors of success in college and the workplace. Students who successfully complete Algebra II as their highest math course in high school are more than five times as likely as students who only complete Algebra I to attain a bachelor's degree. However, few minority students take higher level math courses in high school as shown in Figure 3-7. Asian students outnumber all other groups in taking pre-calculus and calculus.

Reports document a declining student interest in STEM and the fact that too many students are not adequately prepared to succeed in college-level coursework. However, reports consistently show that students who have access to high-level and rigorous coursework and who are taught by teachers with high levels of experience and high expectations for performance are more likely to be prepared for and succeed in the STEM fields regardless of race/ethnicity or socioeconomic status (ACT, 2006; Lleras, 2008).

Dr. Ronald F. Ferguson, a senior lecturer in education and policy at Harvard University Graduate School of Education, expects standardized tests, such as the SAT and ACT, to effectively measure the achievement gap over time. Although more students, especially underrepresented minorities, are taking the SAT, the growth in test takers is reaching deep into the high school student pool and testing lower-achieving students. Others opine that the SAT and ACT are especially poor metrics for measuring trends in the achievement gap because the population of test takers is not stable. These assessments are undertaken only by students who plan to attend college, and the proportion who fall in that group has changed over time. As a result, it is not possible to discern whether changes in the achievement gap reflect changes in performance levels or changes in who is or is not taking the tests. The NAEP 12th grade exam is subject to the same weakness, because high school dropouts are not tested. For this reason, the NAEP 8th grade exam provides a more stable metric with which to judge trends in achievement gaps. NCES surveys such as High School and Beyond (high school class of 1982), the National Educational Longitudinal Study (high school class of 1992), and the Educational Longitudinal Study (high school class of 2004) maintain dropouts in their follow-up samples, so they offer better assessments of the achievement gap at the end of high school than the NAEP or college entry assessments. "The numbers are increasing," Ferguson warns. "We need better instruction and better instruction is going to require better leadership. The fact that the scores aren't going up with the numbers [means that] we have to do more than act on a slogan. We have to prepare students for college" [37]

The Need for Sustained Systemic Intervention and Reform

Federal support for interventions has been agency and program specific, with little cohesion and synergy. Also, the assessments of such interventions tend to document whether they have accomplished their program goals, rather than systemic outcomes. Current measures do not attest to the cumulative impact of these national investments, too few target underrepresented minorities, and there is no systematic way to translate the results of the research into classroom applications.

Partially addressing the issues, the Obama administration has issued *A Blueprint for Reform* to guide the reauthorization of the Elementary and Secondary Education Act (ESEA),[38] replacing the No Child Left Behind Act. It challenges states, districts, and schools to ensure that *all* students graduating or on track to graduate from high school are ready for college and a career by 2020. The priorities include evidence-based rigorous standards to improve performance in high-need schools, equitable distribution of quality teachers and resources, innovative programs for English Learners, and rewards for performance. The blueprint proposes to strengthen formula grant programs for Native American, Native Hawaiian, and Alaska Native education, giving more flexibility to tribal education departments in managing programs and services for Indian students within their jurisdiction.

The blueprint calls on states to provide high-quality STEM instruction by leveraging federal, state, and local funds to integrate evidence-based, effective mathematics or science programs into the teaching of other academic subjects. It emphasizes the need to provide substantial support to high-need schools, including professional development for teachers and school leaders, high-quality curricula, instructional materials and assessments, and interventions that assure that all students are served effectively. Priority will be given to states adopting common, state-developed, college- and career-ready standards.

[37] J. L. Plummer. More diversity among 2009 SAT test takers, scores slightly down, *Diverse Education*, August 26, 2009. Found at http://diverseeducation.com/cache/print.php?articleId=12973.

[38] *A Blueprint for Reform: The Reauthorization of the Elementary and Secondary Education Act.* U.S. Department of Education, March 2010.

4

Access and Motivation

AUGMENTING THE POOL

The science and engineering workforce in the United States—workers with a bachelor's degree or higher in an S&E occupation—is drawn from undergraduates at our nation's postsecondary institutions and immigrants who arrive for graduate study, postdoctorate fellowships, or work in STEM. The base from which we can draw underrepresented minorities for S&E, therefore, is the pool of underrepresented minorities who are enrolled in postsecondary institutions and who plan to complete a four-year degree.

This pool of underrepresented minorities and women enrolled in postsecondary institutions has increased. As shown in Figure 4-1, total undergraduate enrollment across all fields for each racial/ethnic minority group increased between 1976 and 2008. As the National Center for Education Statistics (2010) reports: [1]

- Asians/Pacific Islanders had the fastest rate of increase between 1976 and 2008 (561 percent); their enrollment increased from 169,000 to 1,118,000.
- During the same time period, Hispanic enrollment increased from 353,000 to 2,103,000, a 495 percent increase.

---

[1] National Center for Education Statistics. 2010. *Status and Trends in the Education of Racial and Ethnic Minorities* (NCES 2010-015), July 2010. http://nces.ed.gov/pubs2007/minoritytrends/ (accessed July 15, 2010).

• American Indian/Alaska Native enrollment increased from 70,000 to 176,000, a 151 percent increase.

• Black enrollment increased from 943,000 to 2,269,000, a 140 percent increase.

The enrollment for each minority group rose at a faster rate than that of whites, which increased from 7,740,000 to 10,339,000, a 34 percent increase during this period. The increase in Hispanic enrollment, in particular, reflects a significant growth in the Hispanic segment of the population generally and is projected to continue strong into the future.

Similarly, females of all groups showed an enrollment increase and actually surpassed the percentage of males enrolled as undergraduates in 1980. Black students had the largest difference between male and female enrollments. Black females accounted for 64 percent of the total undergraduate black enrollment. American Indian/Alaska Native females made up 60 percent of the total American Indian/Alaska Native student enrollment, and Hispanic females made up 58 percent of the total Hispanic student enrollment. White females made up 56 percent of the white student enrollment.

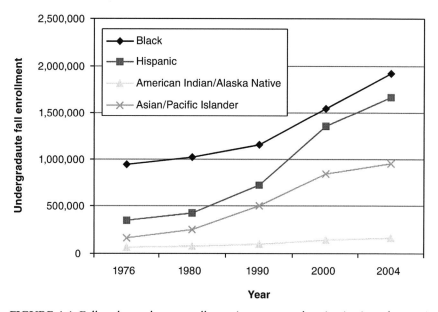

FIGURE 4-1 Fall undergraduate enrollment in postsecondary institutions, by race/ethnicity, 1976-2004.
SOURCE: National Center for Education Statistics, *Status and Trends in the Education of Racial and Ethnic Minorities* (NCES 2007-039), Table A23.1, September 2007.

As a result of these trends over three decades, underrepresented minorities—African Americans, Hispanics, Native Americans, and Alaska Natives—now comprise more than one-quarter (27 percent) of total undergraduates. Though this does fall short of the underrepresented minority proportion in the college age (18- to 24-year old) population (33.2 percent), this is a significant achievement of historical proportions that has its origins in the migration of African Americans to the North during and after World War II, the rapid growth of the U.S. Hispanic population during the past two decades, the civil rights movement of the 1950s and 1960s and its focus on education, and efforts—including affirmative action, financial aid, and institutional efforts to increase diversity—to increase the access of underrepresented minorities and economically disadvantaged students to higher education that date from the 1960s and 1970s.[2]

At the same time that underrepresented minority enrollment has been increasing, the proportion of underrepresented minority freshmen at four-year colleges and universities who aspire to major in STEM fields has increased and, as seen in the HERI data discussed earlier, is and has been since the early 1990s similar to that of whites and Asian Americans. These positive trends provide encouragement that further efforts to stimulate postsecondary enrollment and aspirations to major in STEM should increase the number of underrepresented minorities who are prepared for college, major in STEM, and complete a degree in a STEM field.

There are two important caveats to this picture that indicate areas requiring additional effort. First, the HERI data focus on students at four-year institutions. Underrepresented minorities at two-year institutions, who comprise more than half of all underrepresented minorities enrolled in postsecondary institutions, have a lower propensity to major in and complete degrees in STEM than those who begin at four-year institutions. Second, those underrepresented minorities who do begin at four-year institutions and aspire to major in STEM, as we have seen, have a lower four- and five-year completion rate than whites and Asian Americans.

Clearly preparation matters, as discussed in the previous chapters. Efforts to provide academic, social, and professional support matter as well, as we will discuss in the next chapter. Here we focus on the transition from secondary to postsecondary school, a time when information, motivation, and financial support are critical to aspiring to college enrollment, majoring in STEM, and sustaining interest once enrolled. These efforts can further augment the pool from which we can develop new underrepresented minority scientists and engineers.

---

[2] W. G. Bowen and D. Bok. 1998. *The Shape of the River: Long-Term Consequences of Considering Race in College and University Admissions.* Princeton, NJ: Princeton University Press, pp. 1-14.

ASPIRING TO COLLEGE

Information about options and opportunities in higher education is critical. Improved information can increase the awareness of students and their families so they can best prepare and apply for postsecondary education. Many students, including underrepresented minorities and especially those who are first-generation undergraduates, have insufficient information about educational and career opportunities and options at critical decision points in middle and high school. Most students, in fact, have few opportunities to learn about these options unless institutions—schools, churches, community groups, youth organizations—make an effort to provide role models and information. The Spellings Commission focused in this regard on college awareness activities during high school, noting that "many students and parents don't understand the steps needed to prepare for college," and there need to be resources for early and ongoing college awareness activities, academic support, and college planning and financial aid application assistance."[3] Similarly, a recent report of the College Board argued that, especially to encourage more first-generation students to apply to college, the postsecondary admissions process needs to be simplified and clarified.[4]

The college admissions process is also a critical gatekeeper for postsecondary education. Since the 1950s, the nation's ongoing dialogue and struggle over civil rights has focused squarely on admissions to college, graduate school, and professional programs. This is critically important to underrepresented minority participation in postsecondary education and, by extension to postsecondary STEM, and we cannot place a strong enough emphasis on it. The continued ability of colleges and universities to act affirmatively to ensure inclusion of underrepresented minorities in diverse campus environments will affect the quality of education and the associated opportunities that go with it for these students. In this vein, we must continue to follow experiments in college admissions and assess the differential outcomes of strategies that allow explicit consideration of race or ethnicity in admissions or those that do not—such as those that provide admission to the top 10 percent of a graduating class—but seek a similar goal. Fortunately, the American Association for the Advancement of Science, the Association of American Universities, and the National Action Council for Minorities in Engineering have been closely and carefully following civil rights in academia and the legal cases focused on them through a series of workshops, consultations with institutions, an important report—*Standing*

---

[3] *A Test of Leadership: Charting the Future of U.S. Higher Education.* A report of the commission appointed by Secretary of Education Margaret Spellings (September 2006), p. 18-19.

[4] College Board. 2009. *Coming to Our Senses.* New York, NY: College Board.

Our Ground,[5] and a new handbook—*Navigating a Complex Landscape to Foster Greater Faculty and Student Diversity in Higher Education.*[6] We strongly urge AAAS, AAU, and NACME to continue their efforts and institutions of higher education to participate in the ongoing conversations and consultations they are engaged in.

Finally, underrepresented minority students should be encouraged to attend institutions of higher education that are a solid match for their levels of preparation and motivation. Richard Sander introduced the "mismatch hypothesis" in 2004 which suggests that minority students are less successful as science majors when they are placed in institutions with academic standards that far exceed their preparation. Bowen and Bok (1998) refuted the "mismatch" hypothesis in their analysis of college and beyond data on undergraduates. They found that attending a more selective institution is associated with a higher likelihood of earning a professional or doctoral degree, leads on average to greater career success as measured by annual income, and is, in most instances, correlated with greater involvement (taking leadership positions) in a broad range of civic activities.[7]

The "mismatch" hypothesis has been challenged also by affirmative action advocates and researchers (e.g., Alon and Tienda, 2004) who report that minority students thrive in selective schools despite their disadvantaged starting lines. They conclude that the likelihood of graduation increases as the selectivity of the institution attended rises.

Espenshade and Radford (2009), after extensive research, noted that underrepresented minority students at selective institutions graduate at lower rates than do white and Asian students and end up with grade point averages in the lower ranks of their class. However, they concurred with Bowen and Bok that the advantages associated with attending a more selective institution trump lower class rank (Box 4-1).[8]

A large number of underrepresented minority students "undermatch." As discussed in Box 4-2, William Bowen, Matthew Chingos, and Michael McPherson in their recent examination of college completion found that many African American and Hispanic students attend institutions that are less demanding than they are qualified to attend. This has implications for eventual completion because selectivity of an institution is positively cor-

[5] AAAS & NACME, *Standing our Ground: A Guidebook for STEM Educators in the Post-Michigan Era,* AAAS & NACME, October 2004.

[6] AAAS & AAU, *Navigating a Complex Landscape to Foster Greater Faculty and Student Diversity in Higher Education,* 2010.

[7] W. T. Bowen and Derek Bok. 1998. *The Shape of the River.* Princeton, NJ: Princeton University Press.

[8] T. Espenshade, and A. Radford. 2009. *No Longer Separate, Not Yet Equal: Race and Class in Elite College Admission and Campus Life.* Princeton, NJ: Princeton University Press.

BOX 4-1
No Longer Separate, Not Yet Equal

Although attending a more selective college negatively affects students' relative GPAs when students' ability is held constant, the literature suggests that the benefits in terms of educational attainment, occupational status, and earnings outweigh this downside risk. . . . [T]he advantages of a degree from a more selective college appear to hold for all students, including minority students. On the whole, the evidence does not support assertions made by mismatch proponents. Instead, affirmative action, which enables more underrepresented minority students to gain access to selective colleges than would a race-blind admission policy, appears to help more than harm minority students' futures.

SOURCE: T. Espenshade and A. Radford. 2009. *No Longer Separate, Not Yet Equal: Race and Class in Elite College Admission and Campus Life.* Princeton, NJ: Princeton University Press.

BOX 4-2
Aiming High

Student's choices of where to apply to college are enormously important. A surprisingly large number of students—especially those from poor families and those who are African American or Hispanic—"undermatch." That is, they go to four-year institutions less demanding than those they are qualified to attend, to two-year colleges, or to no college at all. For example, 59 percent of students in the bottom quartile of family income undermatch; 27 percent in the top quartile do so. In addition, 64 percent of students whose parents have no college education undermatch, compared with 41 percent of those whose parents have college degrees and 31 percent whose parents have graduate degrees. Undermatching has serious consequences because there is a strong association between institutional selectivity and BA completion rates: Students with essentially the same qualifications who attend more selective universities have a considerably higher probability of graduating than do comparable students who attend less selective universities. Our data also confirm the results of other studies that show that students whose objective is to earn a BA are much less likely to do so if they start at a two-year college (again, other things equal).

SOURCE: William G. Bowen, Matthew M. Chingos, and Michael S. McPherson, Helping students finish the 4-year run," *Chronicle of Higher Education*, September 8, 2009 (based on Bowen et al. 2009. *Crossing the Finish Line: Completing College at America's Public Universities.* Princeton, NJ: Princeton University Press.

related with completion.[9] At the same time, however, the phenomenon of "overmatching," that is, placing a student in an environment that is too challenging based on previous preparation, is equally a concern. The latter may be addressed by matching a student with a more appropriate institution or by ensuring that, in the more challenging environment, programs are put in place to accelerate the preparation of motivated students. In this latter situation, senior administrators must actively endorse and support minority programs in order to promote faculty buy-in; respected faculty in STEM fields must act as mentors, advisors, role models, and advocates; and the culture of the institution must insist that faculty and others hold everyone to the same high standards and that there be an expectation of success, as measured by both completion and high performance.[10]

MAJORING IN STEM

Academic preparation and admission to a postsecondary institution are the important prerequisites for careers in STEM. Many underrepresented minorities may come from academic backgrounds that, on average, provide less effective preparation for STEM courses than their majority counterparts. Because factors such as the number of years of science and math in high school, high school grades, and standardized test scores in math are positively correlated with choosing to major in science (Maple and Stage 1991; Ware and Lee 1988), many minority students start college less likely to pursue science than their majority classmates. In fact, many of the successful interventions have been designed to address these issues directly by providing opportunities to increase mathematical and analytical abilities (e.g., Treisman 1992; Bonous-Hammarth 2000).

The next step is to encourage an aspiration to major in STEM. Outreach efforts from government agencies, industry, and postsecondary institutions can all work to raise interest and awareness of STEM careers in all students, including underrepresented minorities. NACME has argued that this should begin very early in elementary school, urging that businesses "form partnerships with K-12 schools to promote STEM careers and education to underrepresented minority students, including providing STEM employees to serve as role models and mentors, offering on-site internships to students

---

[9] William G. Bowen, Matthew M. Chingos, and Michael S. McPherson. 2009. "Helping students finish the 4-year run." *Chronicle of Higher Education*, September 8, 2009 (based on Bowen et al., *Crossing the Finish Line: Completing College at America's Public Universities*, Princeton, NJ: Princeton University Press).

[10] R. Tapia, Minority students and research universities: How to overcome the mismatch, *Chronicle of Higher Education* 55(29):A72.

and teachers, and providing access to the latest equipment and software."[11] The same can be said for federal science agencies as well.

Postsecondary institutions have a role in outreach as well. In their evaluation of the NSF's Louis Stokes Alliances for Minority Participation (LSAMP) program, Clewell et al. (2005) describe the kinds of high school outreach activities undertaken by institutions with LSAMP funding:

> More than half of the Alliances also offer high school outreach activities. This includes LSAMP students visiting local high schools to give a science demonstration, tutoring high school students in STEM subjects, helping out at high school science fairs, and disseminating LSAMP recruitment material to high school staff members and students. In some instances LSAMP collaborates in the outreach efforts of other STEM intervention programs that specifically target high school students. Examples include female LSAMP students visiting high schools to talk to girls about math, LSAMP students participating in a precollege initiative where high school students are invited onto the college campus to learn about science disciplines, and science faculty visiting high schools on Saturdays to expose students to science professions and activities.[12]

There are other NSF programs that provide outreach from postsecondary institutions to K-12 schools, including the GK-12 Program, Opportunities for Enhancing Diversity in the Geosciences Program, and Mathematics and Science Partnerships. Some are not targeted directly at underrepresented minority students, but they may benefit them. The National Institutes of Health has additional undergraduate programs focused on underrepresented minorities, such as the Bridges to the Baccalaureate Program, which also includes a high school outreach component.[13]

Mathematics and science summer programs, such as the Upward Bound programs (part of the U.S. Department of Education's TRIO[14] program), provide another means for developing the interest of high school students in these fields. These programs provide an opportunity for students to take summer courses in mathematics and science, engage in research for the first time, and raise awareness of both STEM careers and the steps necessary

---

[11] J. B. Slaughter. 2008. The "new" American dilemma: An open letter from Dr. John Brooks Slaughter, in NACME, *Confronting the "New" American Dilemma: Underrepresented Minorities in Engineering: A Data-Based Look at Diversity*, NACME, p. 8.

[12] B. C. Clewell et al. 2005. *Evaluation of the National Science Foundation Louis Stokes Alliances for Minority Participation Program (Final Report)*. Washington, DC: The Urban Institute, p. 23.

[13] National Research Council. 2005. *Assessment of NIH Minority Research Training Programs: Phase 3*. Washington, DC: The National Academies Press.

[14] TRIO programs are eight federal outreach and student services programs to serve and assist low-income individuals, first-generation college students, and persons with disabilities to progress through the academic pipeline from middle school to postbaccalaureate programs.

along the pathway to them. At present the TRIO Upward Bound Program has been found to be "not performing" by the U.S. Office of Management and Budget, which noted that "Interim findings from an evaluation of the Upward Bound program, released in 2004, indicated that Upward Bound had not been effective in increasing the overall college enrollment rates of its participants."[15] Several members of this study committee personally benefited from working with the Upward Bound program and found it to be a key experience for motivating their interest in pursuing a STEM education in college and beyond. Given the potential benefit of this program, we strongly hope and expect that the U.S Department of Education will take the necessary steps to improve program efficiency and effectiveness.

To complement efforts to raise awareness of STEM careers generally, counseling in middle and high schools can also provide important and timely information in a practical way about what is academically necessary—in high school and in college—to pursue STEM careers. This counseling can also focus on preparing students and families for their initial interactions with higher education institutions, including the application and financial aid processes. Clewell et al. (2005), in reviewing the relevant literature, found:

> There is a great deal of research to establish a strong relationship between career development and student background, particularly socioeconomic status (Hill, Pettus, and Hedin 1990; Mestre and Robinson 1983; Rolle 1977). Scientists tend to come from well-educated white families (Grandy 1994; Pearson 1986). Lack of knowledge and familiarity on the part of underrepresented minorities in terms of what constitutes careers in STEM may contribute to their limited presence in these fields (Hill, Pettus, and Hedin 1990). Knowledge about STEM careers and exposure to scientists and engineers have been found to increase minority students' commitment to a STEM major, degree aspirations, and commitment to a STEM career (Good, Halpin, and Halpin 2001; Rolle 1977; Wyer 2001).[16]

Unfortunately, academic and career counseling is often weak in predominantly minority secondary schools. Moreover, it can be counterproductive, steering minority students into less demanding courses and programs when they should be challenging students by encouraging them to take the highest level courses they are prepared for.

---

[15] http://www.whitehouse.gov/omb/expectmore/summary/10000210.2002.html (accessed February 22, 2010).

[16] B. C. Clewell et al. 2005. *Evaluation of the National Science Foundation Louis Stokes Alliances for Minority Participation Program* (Final Report). Washington, DC: The Urban Institute, p. 39.

SUSTAINING INTEREST AND MOTIVATION

Providing information and creating awareness about STEM education and careers are critical, but generating real motivation to pursue a STEM career requires something more. The kinds of research experiences embedded in summer programs, such as the Upward Bound program, are one way to ignite a passion for science. Another way is to appeal to the personal interests of students and their families. As shown in Box 4-3, a strong case can be made to all students that they can pursue a science career and give back to their families and communities at the same time.

Willie Pearson found, in his study of African American PhD chemists, that reading the biographies or biographical sketches of eminent African American scientists influenced a number of the chemists he interviewed to pursue science careers. The biographies were typically published in minority-focused magazines or reference books. Pearson contends that most American historians and sociologists have largely ignored African American scientists in general and chemists in particular. He recommends that historians and sociologists of science research and publicize—in school curricula and popular magazines—the contributions and experiences of eminent African American scientists.[17]

REPRISE ON FOUR APPROACHES

Reflecting for a moment on the four approaches to improving underrepresented minority participation in STEM presented in chapter 3 and in particular in Table 3-1, we can see each of these approaches at play here:

*   *Improving Information for All Prospective College Students*: The Spellings Commission recommendation to improve college awareness activities, aimed at all students, is important to and can benefit underrepresented minorities as well as others, including those in STEM.
*   *Increasing the Pool of Undergraduate Underrepresented Minorities*: Efforts to ensure that underrepresented minorities have a fair chance at admission to a postsecondary institution they are qualified for have included policies ranging from affirmative action to admissions policies that offer automatic admission of top students to state institutions. Such policies are fundamentally important to increasing the participation of underrepresented minorities at the postsecondary level across all fields, including those in STEM as evidenced by the focus of AAAS and NACME on this issue.

---

[17] W. Pearson. 2005. *Beyond Small Numbers: Voices of African American PhD Chemists.* Stamford, CT: Jai Press, pp. 150-151.

BOX 4-3
Why African American Students Should Major in
Biomedical Research

There are statistics that are profoundly disturbing. About thirty years ago, an NSF report showed that less than 1 percent of the PhD degrees in science, engineering, and mathematics fields in the US were awarded to African Americans. Substantial amounts of money were spent by the federal government over the past 30 years to address this problem, yet as of last year, the number of PhDs in STEM fields awarded to African Americans climbed to about 2 percent.

Why should this matter to you?

I would like to share with you some information that I received at a recent meeting at the NIH that focused on minority health issues. Before I do, I would like for you to take a careful look at your mother, if you came here with your mother. Seriously. Now, take a look at your father.

- If your father is African American, his risk of dying of high blood pressure is 350 percent greater than that of his white friends.
- If your mother is African American, her risk is 300 percent greater.
- If your parents are African American, their risk of developing hypertension is 200 percent greater than the risk of Caucasians.
- If your mother is African American, there is a 1 in 4 chance that she will develop diabetes by the time she is 55.
- African American women have a 3-fold greater chance of developing Lupus.
- Consider also the fact that sickle cells were discovered near the turn of the century, but it was nearly 50 years before the first penny of federal funds were spent to study the disease.

Who do you expect to address these issues? Who will do the research, or make funding decisions? You will be among the brightest students in your class when you go to college, and you will have the opportunity to do more than just put band-aids on problems. You will have the opportunity to find new cures.

SOURCE: Michael Summers, Remarks to High-Achieving Minority High School Students, University of Maryland Baltimore County, 2008.

• *Raising Awareness of STEM Careers*: There is general concern about the participation of U.S. citizen students in STEM fields, regardless of race and ethnicity. One set of strategies for addressing this includes K-12 awareness activities, improved counseling for science and mathematics, and activities that promote STEM (e.g., the FIRST Robotics Competition). If these are made universally available, they will benefit underrepresented minorities as well as others.

- *Increasing STEM Outreach to Underrepresented Minorities*: Programs such as the LSAMP high school outreach activities and the TRIO Upward Bound Program that specifically target underrepresented minorities in mathematics, science, and engineering are important means for reaching these groups and providing a pathway forward in STEM.

5

Affordability

The availability of financial support affects postsecondary attendance and persistence for students in general and for underrepresented minority students in particular. This should come as little or no surprise, especially in this era of rising tuition; it should also be no surprise that the participation of underrepresented minorities in STEM is affected by trends in tuition and aid in many of the same ways as other students are affected. However, there are issues that are specific to STEM and others specific to underrepresented minorities. Further, there are important differences between aid at the undergraduate and graduate levels. But in the end, however one slices it, money matters.

FINANCIAL SUPPORT AND COLLEGE COMPLETION

Clewell et al.(2005) found, when reviewing the literature, that financial support has been demonstrated to have a "positive influence on student persistence" (Murdock 1987, St. John 1991, St. John, Kirschstein, and Noell 1991).[1] In the most recent study of college completion, Bowen et al. (2009) reported:

> We find big gaps by family income in completion rates and in the time it takes to earn degrees—even after we control for related differences in

---

[1] B. C. Clewell et al. 2005. *Evaluation of the National Science Foundation Louis Stokes Alliances for Minority Participation Program (Final Report)*. Washington, DC: The Urban Institute, November, p. 38.

factors like parental education. For example, at the flagships 83 percent of students from the top half of the income distribution graduate within six years, but only 68 percent from the bottom half do so: a difference of 15 percentage points. The difference in four-year graduation rates is 19 points. We also find that differences across states in the net prices paid by students have significant effects on the odds that a low-income student will graduate: the higher the net price, the lower the completion rate (other things equal). On the other hand, there is no correlation between net price and completion rates for high-income students, a finding that raises real questions about the wisdom of merit-aid programs and policies aimed at keeping tuition low across the board.[2]

Because of the importance of financial aid to college attendance and completion, the College Board has recently argued that it is important to keep college affordable "by controlling college costs, using available aid and resources wisely and insisting that state governments meet their obligations for funding higher education." More specifically, the College Board recommended "more need-based grant aid while simplifying and making financial aid processes more transparent." The College Board (2009) noted that need-based aid should keep pace with inflation; student debt should be minimized; financial aid processes should be made more transparent and predictable; and institutions should be given incentives to enroll and graduate more low-income and first-generation students.[3]

Indeed, one of the most compelling factors affecting the supply of minority STEM graduates has involved financial incentives and the availability of targeted scholarships. Yet this has been one of the most highly debated and legally attacked issues in higher education. The highly visible federal court actions (*Regents of the University of California v. Bakke, Hopwood v. Texas, Johnson v. Board of Regents,* and *Gratz v. Bollinger*) mainly addressed race in admissions decisions.

> While critically important for those selective institutions that consider race as part of the admissions process, the affirmative action issue in financial aid has significance—and potential impact—that extends beyond the question of admissions. First, minority students are more likely to come from low-income families. As a result, for most of these students, the availability of financial aid is a significant factor affecting their ability to go to college. Second, at a time of increasing national diversity, and with the recognition that we can "leave no child behind," we face the prospect that by not providing the necessary financial aid supporting college and university

---

[2] William Bowen, Matthew C. Chingos, and Michael S. McPherson. Helping Students Finish the 4-Year Run. *The Chronicle of Higher Education.* September 8, 2009. Available at http://chronicle.com/article/Helping-Students-Finish-the/48329.

[3] College Board. 2009. *Coming to Our Senses.* New York, NY: College Board.

attendance, college campuses "will be missing 800,000 otherwise qualified minority students between now and 2015, with the commensurate losses of billions of dollars to the national economy.[4]

Race-based scholarships and fellowships funded by federal agencies and states have been subject to widespread reform as the result of legal challenges. For example, in *Podberesky v. Kirwan* (1994), the United States Court of Appeals for the Fourth Circuit held that a race-exclusive merit scholarship program at the University of Maryland at College Park was unconstitutional. The Benjamin Banneker scholarship for African Americans was consolidated with another campus-based program. Similarly, the Minority Graduate Research Fellowship Program administered by the National Science Foundation was eliminated as a separate program following a lawsuit challenging that it was discriminatory. Despite these challenges and the continuing dialogue about the effectiveness of race-neutral policies, no one denies the fact that we had the most rapid growth of minorities in STEM fields during this period.

COLLEGE AFFORDABILITY

College affordability has been a perennial issue since World War II, particularly for low- and middle-income students. Affordability and opportunity were first addressed by the Serviceman's Readjustment Act of 1944 (commonly known as the G.I. Bill), which dramatically increased both college enrollment and the size of the American middle class, though research has shown that the benefits were largely for white males.[5] Since then, key questions of support have evolved over time. Today they focus on appropriate mechanisms of support (grants, loans, tax benefits), whether grants should be need based or merit based, and the overall cost of college today. Critics have questioned the increased "costs" associated with undergraduate education, particularly at private colleges, but also among the public ones. Administrators have pointed out, though, that while there have been increases in costs (e.g., information technology, health care) and decreases in state appropriations, much of what appears to be increases in cost has been an increase in tuition combined with an increase in aid, essentially a

---

[4] A. Coleman. 2002. "Diversity in Higher Education: A Continuing Agenda," *Rights at Risk: Equality in an Age of Terrorism*, Report of the Citizens' Commission on Civil Rights. Washington, DC, p. 73.

[5] Sarah Turner and John Bound. 2003. Closing the Gap or Widening the Divide: The Effects of the G.I. Bill and World War II on the Educational Outcome of Black Americans. *Journal of Economic History*. Available at http://en.wikipedia.org/w/index.php?title=African_Americans_and_the_G.I._Bill&printabl.

shift in the college pricing structure from a low-tuition/low-aid approach to a high-tuition/high-aid financial model.

Therefore, the focus needs to be how to allocate aid so that "net tuition" (tuition/fees/room and board minus financial aid: what a student or family will actually pay) is appropriate to the student, family, and institution.

The issue of aid and affordability, though, is complex, because it is affected by state policies, federal programs, and institutional aid. As shown in Figure 5-1, federal aid is the largest source of financial support for undergraduates although, by itself, just about half of all aid. The rest is provided through state and institutional aid.

State governments play a key financial role for public institutions, where the large majority of students are enrolled, and they have two key policy levers with regard to college affordability: state appropriations to institutions (which affect tuition) and individual financial aid programs (which then affect net tuition). Tuition, of course, has risen significantly in the last quarter century, regularly outpacing inflation, but this trend has been substantially affected by trends in state funding for higher education. On a per capita basis, that funding has steadily declined during this period, reaching a 25-year low in 2004-2005 before turning up and then plummeting down again in the current economic recession. Consequently, tuition at public institutions has increased substantially. Institutions often use tuition income to replace revenue deficits, with the concomitant effect of decreasing affordability for all students but especially for underrepresented students from low-income backgrounds.

Meanwhile, as tuition has increased, financial aid has taken on a more salient role. Under the high-tuition/high-aid model, aid that is generally need based will result in a financial model in which wealthier families subsidize those of lower income. However, there has been an ongoing shift at the state level recently from need-based aid toward merit-based financial aid that undermines this model. Merit-based recipients, selected on the basis of test scores, grade-point average, and other academic achievements, accounted for 24 percent of state grants in 2004-2005, up from 9 percent in 1984.[6] Several states have introduced new, academically based aid programs that adopt Georgia's "Helping Outstanding Pupils Educationally" (HOPE) Scholarship program model. Approximately 16 states have implemented such programs with varying qualifying criteria.

Advocates for merit-based programs contend that low-income students are not excluded and that these programs motivate more students to excel academically. Critics, on the other hand, argue that low-income students are disadvantaged because they attend schools that do not have the resources to support academic excellence, and so these students do not have the test

---

[6] IHEP. 2006. Convergence: Trends Threatening to Narrow College Opportunity in America (2006). Institute for Higher Education Policy.

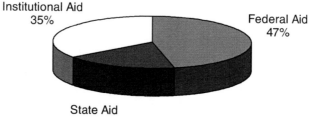

FIGURE 5-1 Source of financial aid received by undergraduates, 2007-2008. SOURCE: U.S. Department of Education, National Center for Education Statistics, 2007-2008 National Postsecondary Student Aid Study (NPSAS:08).

scores and grade-point averages to qualify. Consequently, scholarships from merit-based programs disproportionately are awarded to white and upper-income students. However, for states that have had enough data and history to allow an in-depth analysis of the effects by race (Arkansas, Florida, and Mississippi), Dynarski (2004) found that merit-based programs have helped to close racial gaps in participation.[7] This may be attributed to the simplicity of the programs themselves and the wide publicity given to these programs among high school counselors. However, additional research on a broader range of states is needed to provide a clearer view of the effects by race/ethnicity of state shifts to merit-based aid.

The **federal government** plays a dual role that focuses both on college access generally and support for students in STEM programs more specifically. Funding for federal financial aid programs that are primarily need-based have increased over time: for example, Pell Grants, Supplemental Educational Opportunity Grants (SEOG), Perkins loans, and subsidized Stafford loans. Unfortunately, the maximum Pell Grant award, the largest direct college subsidy, has not increased proportionately to tuition increases and inflation-adjusted dollars, and newly implemented eligibility policies are expected to shift the income threshold and exclude students who currently qualify for Pell Grants. The current administration has, in the meantime, proposed program changes that would increase the maximum Pell Grant by $200, to $5,550, for the 2010-2011 academic year. If this is enacted, an additional 260,000 students would be eligible for a grant. Further, the administration has proposed making the Pell Grant an entitlement. In so doing, the administration would index the maximum award to the Consumer Price Index (CPI) plus a percentage point in order to help shift the

---

[7] Ibid.

rate of increase relative to tuition increases. That change, if enacted, would take effect in 2010-2011.[8]

While funding for need-based programs has not kept pace with inflation, funding for programs that do not target low-income students—unsubsidized Stafford loans, federal loans to parents (PLUS), and tax benefits have increased at faster rates. Thus, federal policy—similar to state policies—has also shifted support toward middle- and upper-income students. Much of this redirection of federal aid is the result of pressure from middle-income families to make college more affordable. The focus is on affordability rather than access.

In addition, the use of **loans** as part of undergraduate financing plays a further complex role. For example, researchers have found that minorities and students with poor academic preparation have a significant aversion to debt due to the greater risk of loan default.[9] Moreover, even if students from low- and middle-income families recognize the long-term value of an investment in higher education, the debt burden quickly mounts. They may deal with this by leaving school after completion of a bachelor's degree, reducing the number of minority students continuing to graduate school. Or they may deal with it by working full- or part-time, which disadvantages them relative to others because they cannot concentrate full time on their studies and research. The effect of these financial factors is seen in persistence rates and degree attainment. For example, for the 2003 cohort who started at four-year institutions, 73 percent of African Americans and 76 percent of Hispanics were still enrolled or with a certificate/degree three years after enrolling, compared to 83 percent of whites and 89 percent of Asian Americans.[10]

Institutions are pivotal in the recruitment and retention of underrepresented minorities, and they impact the persistence of these students through the provision of institutional aid, much of which is need based. They award 42 percent of all grant aid to undergraduates, whereas the federal government provides 31 percent of the total. In 2006-2007, 80 percent of institutional grant aid was need based in private doctorate-granting institutions with tuition above the median, compared to 61 percent in doctorate-granting institutions with lower tuition.[11]

---

[8] K. Field, Obama's Pell Grant proposal would make 260,000 more students eligible, report says, *Chronicle of Higher Education*, News Blog, March 26, 2009.

[9] A. Dowd and T. Coury, The effect of loans on the persistence and attainment of community college students [Electronic version], *Research in Higher Education* 47:33-62.

[10] M. Ryu. 2008. *Minorities in Higher Education*. Washington, DC: American Council on Education.

[11] College Board. *Trends in Student Aid: 2008*, Trends in Higher Education Series.

Researchers such as Gross, Hossler, and Ziskin(2007)[12] and Gansemer-Topf and Schuh(2005)[13] have examined the relationship between institutional aid and student persistence. They cite the need for institutions, especially low-selectivity institutions, to allocate more resources for institutional scholarships and grants in order to improve retention and graduation. High-selectivity institutions should divert resources for the same purpose also, if they are committed to diversifying the student body by socioeconomic status or other demographics.

Some institutions have done precisely that. For example, Harvard University expanded financial aid for low- and middle-income families by eliminating the requirement for parents in families with less than $60,000 income to contribute to the cost of their children attending the institution. Harvard also reduced the contributions of families with incomes between $60,000 and $80,000. Similarly, Brown University approved a new financial aid policy that eliminates loans for students whose family incomes are less than $100,000, reduces loans for all students who receive financial aid, and no longer requires a parental contribution from most families with incomes of up to $60,000. The G. Wayne Clough Georgia Tech Promise Program offers financial awards to Georgia residents whose families have an annual income of less than $33,300 by filling a gap in the financial aid support system for these students.

Some authors reference Tinto's (1993) interactionalist theory of student departure as the basis for their hypotheses about the impact of institutional aid. Tinto proposed that the more students interact with their academic and social environments, the more likely they are to persist. He acknowledged that organizational behavior is an important way to enhance a student's integration to his or her institution. Institutional financial aid can be viewed as an expression of commitment.

In addition to these more global policies, institutions can often be helpful in an ad hoc, just-in-time manner by providing small amounts of funding. Very often lower-income students struggle with cash flow problems. These are temporary problems, not very large scale, that can be reasonably addressed through an emergency or revolving loan fund. The University of Texas El Paso, for example, addresses these issues in two ways: First, there is a revolving loan fund for the purchase of books. Students can take out a loan at the beginning of the semester to pay for books and then pay off the loan over the course of the semester. Second, the university has an emergency loan fund that can help students in need

---

[12] J. P. K. Gross, D. Hossler, and M. Ziskin. 2007. Institutional aid and student persistence: An analysis of the effects of institutional financial aid at public four-year institutions, *NASFAA Journal of Student Financial Aid* 37(1):28-39.

[13] A. M. Gansemer-Topf and J. H. Schuh. 2005. Institutional grants: Investing in student retention and graduation. *NASFAA Journal of Student Financial Aid* 35(3):5-20.

deal with a specific issue. Unfortunately, small cash flow problems and emergencies that have financial implications can derail a student's education, and they need not do so. A small infusion of funds can help tide over students and keep them on track.

In conclusion, we note two recent reports that have focused in full or in part on the subject of financial aid and its potential reform.

First, the Rethinking Student Aid study group convened by the College Board (2008a) recently recommended a major overhaul of the financial aid system.[14] Specifically, they propose the following:

- Make federal financial aid simple, clear, and transparent
- Target loan subsidies toward assisting students in repayment
- Develop a savings program for low-income families analogous to the current federal savings programs that subsidize the college savings of wealthier families
- Provide incentives that reward colleges and universities for supporting their students successfully through college toward completion of their degrees and incentives for states to support the goals of the federal aid system.

Second, the Spellings Commission also sought to improve financial aid by recommending an improvement in the financial aid process, transparency in net price, and better targeting of financial aid.[15] Specifically, its recommendations included:

- Replacing FAFSA with a shorter and simpler application form;
- Significantly increasing need-based student aid;
- Attending to the financial aid needs of transfer students;
- Consolidating federal grant programs to increase the purchasing power of the Pell Grant;
- Developing, at the institutional level, new and innovative means to control costs, improve productivity, and increase the supply of higher education;
- Making available data on costs and price accessible to consumers; and
- The preparation, by NCES, of timely annual public reports on college revenues and expenditures, including analysis of major changes from year to year, at the sector and state level.

---

[14] College Board. 2008. *Fulfilling the Commitment: Recommendations for Reforming Federal Student Aid in Brief*. The report from the Rethinking Student Aid Group.

[15] A Test of Leadership: Charting the Future of Higher Education. A report of the Spellings Commission. Ibid. p. 19.

There will continue to be dialogue on these issues, to be sure, but these policies regarding support provide the foundation of opportunity for under-represented minority students, as they do for all.

FINANCIAL SUPPORT IN STEM PROGRAMS

The Role of Financial Support

While general need-based and merit-based support are provided by the federal government, states, and institutions, financial support for students in STEM is provided primarily by the federal government with some additional foundation support.

The need for financial support for students in STEM fields has been demonstrated in a series of reports. NCES (2000) found that degree completion in science and engineering was positively related to receiving financial aid.[16] The NRC (2005), in its assessment of the minority research training programs at NIH, found that funding was critical to the success of students in biomedical and behavioral programs. Merit-based financial support allowed these students to focus their time and effort on their studies and research, contributing strongly to their success. Indeed, the report found that when such support is lacking and undergraduates already greatly challenged by a demanding research program in addition to a full course load take on additional outside work to make ends meet, it is a "recipe for disaster" (2005, 8).[17] NCES (2000) noted that parental financial support allows higher-income students to focus on their studies, while ACE (2005) found that the number of hours worked while in college was strongly related to persistence in STEM—noncompleters were more likely to have been working 15 hours or more per week—and Oseguera et al. (2006) found that the need to work during the undergraduate years can complicate the pursuit of majors perceived to be time-intensive, including those in the sciences.[18]

At the graduate level, the Council of Graduate Schools has found through their PhD Completions Project that financial support, mentoring/advising, and family support are the main factors that contributed to the completion of doctoral degrees. CGS reports that four-fifths (80 percent) of respondents indicated that financial support was a main factor in their ability to complete their doctoral program. Graduates from mathematics and physical sciences programs were the most likely to report that financial

---

[16] National Center for Education Statistics, *Entry and Persistence of Women and Minorities in College Science and Engineering Education* (NCES 2000-601). Washington, DC: U.S. Department of Education, 2000.

[17] National Research Council. 2005. *Assessment of NIH Minority Research Training Programs; Phase 3*. Washington, DC: The National Academies Press, p. 8.

[18] Oseguera et al. 2006.

support was one of the main factors enabling them to complete their degree (83 percent), followed by engineering and life sciences (both at 82 percent, social sciences (80 percent), and humanities (73 percent).[19]

Financial Support for Undergraduates in STEM

Rising Above the Gathering Storm argued that the educational attainment of U.S. students in the natural sciences and engineering lags behind that of other OECD countries and recommended national action to address the gap so that we can sustain our competitiveness in a global economy that requires high wages to be justified by talent. It recommended that the United States "increase the number and proportion of U.S. citizens who earn bachelor's degrees in the physical sciences, life sciences, engineering, and mathematics by providing 25,000 new 4-year competitive undergraduate scholarships each year to U.S. citizens attending U.S. institutions.[20]

The Higher Education Reconciliation Act of 2006, which became effective July 1, 2006, created the Academic Competitiveness Grant (ACG) Program and National Science and Mathematics Access to Retain Talent Grant (National SMART Grant) Program, administered by the U.S. Department of Education, partly fulfilling the *Gathering Storm* recommendation. The program awards need-based Academic Competitiveness Grants to first- and second-year undergraduates who have completed a rigorous high school curriculum and National SMART Grants to third- and fourth-year undergraduates majoring in certain technical fields or foreign languages deemed vital to national security. An early audit of the program found that participation in these programs was low and that the Department of Education was not undertaking enough effort to promote the grants.[21] It is our understanding that participation has increased, but with resources more limited, we hope that the current departmental administration will make every effort to ensure that these funds are utilized effectively.

Other federal programs that support undergraduates in STEM include programs administered by the National Science Foundation. NSF STEM education programs that include financial support for undergraduates include:

- The Science, Technology, Engineering, and Mathematics Talent Expansion Program (STEP), which seeks to increase the number of students (U.S. citizens or permanent residents) receiving associate or baccalaureate

---

[19] Council of Graduate Schools. Ph.D. Completion and Attrition: Findings from Exit Surveys of Ph.D. Completers (Released September 2009). Washington, DC.

[20] NAS, NAE, and IOM. 2007. *Rising Above the Gathering Storm*, p. 9.

[21] "Education Dept. Blamed for Not Doing Enough to Promote Grants." *Chronicle of Higher Education*, August 4, 2008.

degrees in established or emerging STEM fields. Financial incentives are provided to students through grants awarded to single institutions and consortia.[22]

- NSF Scholarships in Science, Technology, Engineering, and Mathematics (S-STEM), which makes grants to institutions of higher education to support scholarships for academically talented, financially needy students, enabling them to enter the workforce following completion of an associate, baccalaureate, or graduate degree in science and engineering disciplines. Grantee institutions are responsible for selecting scholarship recipients, reporting demographic information about student scholars, and managing the S-STEM project at the institution.[23]

- Federal Cyber Service, which is a Scholarship for Service (SFS) program that provides funding to institutions to award scholarships in information assurance and computer security fields. Scholarship recipients become part of the Federal Cyber Service of information technology specialists who ensure the protection of the U.S. government's information infrastructure.[24]

In addition to these programs, there are NSF, NIH, and NASA programs that provide financial support more specifically to underrepresented minorities in STEM.

Financial Support for Graduate Education

Data on the financial support for science and engineering graduate students provide two windows into how students are supported in these fields. First, data on graduate enrollment indicate how graduate students—both at the master's and doctoral level—are financed. Second, data on new doctorates provide a picture of how those who complete S&E doctoral degrees were supported.

Enrolled Graduate Students

The data on current S&E graduate students show that for two-thirds of students, their primary financial support came from the federal government, state government, university sources, employers, nonprofit organizations, and foreign government. One-third of current S&E graduate students are self-supporting, relying on personal or family funds, making self-support their largest primary source of support. (This is substantially higher than for S&E doctorate recipients, with just 10 percent reporting personal funds

---

[22] http://www.nsf.gov/funding/pgm_summ.jsp?pims_id=5488 (accessed February 24, 2010).

[23] http://www.nsf.gov/funding/pgm_summ.jsp?pims_id=5257 (accessed February 24, 2010).

[24] http://www.nsf.gov/od/lpa/news/publicat/nsf04009/ehr/due.htm (accessed July 29, 2010).

as the primary source of support in graduate school.) The second largest mechanism is the research assistantship (25 percent), teaching assistant-ships (18 percent), and fellowships or traineeships (12 percent). The federal government is the second largest source, providing financial support for one-fifth of full-time graduate students in 2006.

As shown in Figure 5-2, there is variation by field. For example, in fall 2006, full-time students in physical sciences were financially supported mainly through federally funded research assistantships (RAs) (42 percent) and teaching assistantships (TAs) (38 percent). RAs also were important in agricultural sciences (57 percent); biological sciences (42 percent); earth, atmospheric, and ocean sciences (41 percent); and engineering (40 percent). In mathematics, more than half (53 percent) of full-time students were supported primarily through TAs and another 21 percent were self-sup-ported. Full-time students in the social and behavioral sciences were mainly self-supporting (46 percent) or received TAs (20 percent), and students in medical/other life sciences were mainly self-supporting (60 percent).[25] These variances can be seen in the doctoral data as well.

Financial Support for S&E Doctorates

The federal government is a more significant funder of doctoral educa-tion in science and engineering, primarily through federally funded RAs, but also through a limited number of TAs, individual fellowships, and institu-tional grants that support traineeships. As shown in Table 5-1, the primary source of support for 2007 S&E doctorate recipients was a research assis-tantship (34.5 percent), followed by a fellowship or traineeship 19.4 per-cent), teaching assistantship (14.5 percent), personal support (10.4 percent), and grant/stipend (6.2 percent) (National Science Board, 2010, Appendix Table 2-24).[26] (Box 5-1 also reports data on financial support collected by the Council of Graduate Schools Doctoral Completions project.)

Graduate research assistantships are generally funded through federal research grants awarded to universities. The other primary sources of federal support, particularly fellowships and traineeships, are provided through such programs as:[27]

- Ruth L. Kirschstein National Research Service Award (NRSA) Program, National Institutes of Health ($761.0 million)

---

[25] http://www.nsf.gov/od/lpa/news/publicat/nsf04009/ehr/due.htm (accessed July 29, 2010).

[26] NSB, *S&E Indicators 2010*, Appendix Table 2-24.

[27] Appropriations levels are for FY 2006 and found in Academic Competitiveness Council, Final Report (2007). Appropriations in the ARRA were $12.5 million for the new graduate fellowship program at DOE and were $15.0 for the new Science Master's program at the National Science Foundation.

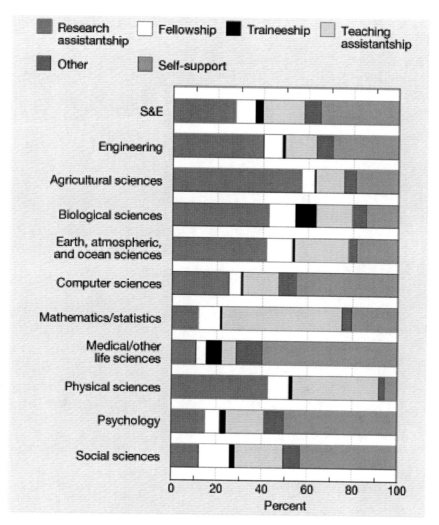

FIGURE 5-2 Full-time S&E graduate students by field and mechanism of support, 2006.
SOURCE: Science and Engineering Indicators 2010, Figure 2-9.

- Graduate Research Fellowships, National Science Foundation, ($93.36 million)
- Integrated Graduate Education and Research Traineeships (IGERT), National Science Foundation ($65.42 million)
- Graduate Teaching Fellowships in K-12 Education, National Science Foundation (GK12) ($50.65 million)

TABLE 5-1 Primary Support Mechanisms for S&E Doctorate Recipients, by Citiz

Support Mechanism	All Doctorate Recipients	Temporary Residents	Unknown Citizenship
All mechanisms			
Number	33,826	12,755	2,408
Percent	100.0	100.0	100.0
Research assistantships[a]			
Number	11,666	6,842	98
Percent	34.5	53.6	4.1
Fellowships or traineeships			
Number	6,566	1,662	45
Percent	19.4	13.0	1.9
Teaching assistantships			
Number	4,907	2,305	35
Percent	14.5	18.1	1.5
Grant/stipend			
Number	2,089	446	12
Percent	6.2	3.5	0.5
Personal[b]			
Number	3,503	301	30
Percent	10.4	2.4	1.2
Other[c]			
Number	1,210	423	10
Percent	3.6	3.3	0.4
Unknown			
Number	3,885	776	2,178
Percent	11.5	6.1	90.4

[a] Research assistantships and other assistantships.
[b] Personal savings, other personal earnings in graduate school, other family earnings or savings, loans.
[c] Employer reimbursement or assistance, foreign support, and other sources.
NOTES: *S&E* includes health fields (i.e., medical sciences and other life sciences). *Total* includes

- Graduate Assistance in Areas of National Need (GAANN) Program, U.S. Department of Education ($32.175 million)
- National Defense Science and Engineering Graduate Fellowships, U.S. Department of Defense ($31.6 million)
- Graduate Fellowships in Science, Mathematics and Engineering, U.S Department of Energy (new)
- Science Master's Program, National Science Foundation (new)

There are other programs as well, though these are the largest.

Sex, and Race/Ethnicity, 2007

	U.S. Citizens and Permanent Residents						
Total	Male	Female	White	Asian	Underrepresented Minority	Other/Unknown Race/Ethnicity
18,663	9,793	8,869	14,178	1,822	2,361	302
100.0	100.0	100.0	100.0	100.0	100.0	100.0
4,726	2,830	1,896	3,670	586	400	70
25.3	28.9	21.4	25.9	32.2	16.9	23.2
4,859	2,436	2,423	3,460	502	829	68
26.0	24.9	27.3	24.4	27.6	35.1	22.5
2,567	1,394	1,173	2,091	212	222	42
13.8	14.2	13.2	14.7	11.6	9.4	13.9
1,631	856	775	1,268	166	184	13
8.7	8.7	8.7	8.9	9.1	7.8	4.3
3,172	1,282	1,890	2,461	183	485	43
17.0	13.1	21.3	17.4	10.0	20.5	14.2
777	512	265	616	75	74	12
4.2	5.2	3.0	4.3	4.1	3.1	4.0
931	483	447	612	98	167	54
5.0	4.9	5.0	4.3	5.4	7.1	17.9

unknown sex. *Underrepresented minority* includes blacks, Hispanics, American Indians/Alaska Natives, Native Hawaiians/Other Pacific Islanders, and multiple races/ethnicities. Traineeships include internship and residency.
SOURCE: NSB, *Science and Engineering Indicators 2010, Appendix Table 2-24.*

While most science and engineering graduate students actually rely on multiple sources of support rather than one primary source, the key to retention and reduced time-to-degree is sustained funding. The ideal funding package—particularly at the graduate school level—would allow the student to focus on studies and research full time, without increasing debt burden, and would include stipend, full tuition and fees, research and travel allowance, cost of living subsidy, health insurance, and other applicable costs of education.

BOX 5-1
Financial Support of Doctoral Completion

The overwhelming majority of respondents received financial support for their doctoral study (94 percent) and 70 percent reported that they were guaranteed multiyear support at the time of admission. Compared with graduates in other fields, engineering and humanities graduates were less likely to report that they had been offered guaranteed multiyear funding at time of admission. For example, only 63 percent of engineering doctoral students and 66 percent of humanities students reported being offered guaranteed multiyear funding at time of admission compared with 72-73 percent of students in mathematics & physical sciences and social sciences and 77 percent of those in life sciences.

Doctoral students in mathematics and physical sciences appeared to have the most generous offers at time of admission, with 22 percent reporting that their offer included six or more years of guaranteed funding, and only 13 percent reporting that they had been offered funding for two to three years. In contrast, only 2 percent of students in social sciences and 8 percent of those in humanities reported receiving offers guaranteeing support for six or more years, and 25-28 percent reported funding for two to three years.

Teaching assistantships tended to be more common in humanities, social sciences, and mathematics & physical sciences than in engineering and life sciences (72-81 percent compared with 51-56 percent). Among those with teaching assistantships, there appeared to be considerable consensus that being a teaching assistant increased the length of the program, and this was particularly true among engineering and life sciences graduates, 88-89 percent of whom reported that teaching assistantships had increased the length of the program.

Close to four-fifths (78 percent) of engineering graduates had received a research assistantship compared with only 28 percent of humanities graduates and 45 percent of life sciences graduates. Among those with research assistantships, there was considerable diversity of opinion regarding its effect on the length of time to degree completion. While 52-54 percent of social sciences and humanities graduates reported that this type of assistantship increased the length of the program, only 22-25 percent of those in life sciences and mathematics & physical sciences fields did so.

Only 60 percent of humanities graduates reported being satisfied with the level of financial support they received during their doctoral program compared with 74 percent of social sciences graduates and 80-85 percent of those in the mathematics & physical sciences, engineering, and life sciences. This is partly explained by the fact that Humanities students were the most likely to work outside the university during their program, to take out loans to support their study, and to report heavier overall burden of debt.

SOURCE: *Ph.D. Completion and Attrition: Findings from the Exit Survey of Ph.D. Completers*, Ph.D. Completion Project. 2009. Washington, DC: Council of Graduate Schools.

A final financial consideration at the graduate level is the availability of funding for professional development activities. To the extent that students can participate in conferences, present papers, engage in summer research, or take advantage of similar activities, the deeper their commitment to their program, their discipline, and their profession. Students from disadvantaged backgrounds will likely require additional financial support for these activities as well. Sources of this support may include institutional funds or funding from federal or philanthropic programs.

SUPPORT FOR UNDERREPRESENTED MINORITIES IN STEM

Returning once again to the four approaches to increasing the participation of underrepresented minorities in STEM, it is clear that underrepresented minorities may benefit from:

- *General financial aid programs*: Need-based programs such as the Pell Grant can and should be used to support low-income underrepresented minority students, including those interested in STEM, in attending college.
- *General programs to assist underrepresented minorities*: Programs designed to support underrepresented minorities in undergraduate or graduate programs, such as the Ford Foundation Fellowship, can be used to support students in STEM as well as other fields.
- *Programs supporting STEM education*: Underrepresented minorities can and should be supported under programs designed to increase U.S. citizen participation in STEM, including the American Competitiveness and SMART Grant programs at the undergraduate level and federal research assistantship, fellowship, and traineeships programs at the graduate level.
- *Programs supporting underrepresented minorities in STEM*: To meet the specific needs of underrepresented minorities in STEM who are not covered by the above programs or who need extra incentives to participate in STEM, additional programs focusing on underrepresented minorities are also important to achieving the national goal of increased participation.

It is to these latter programs we turn now.

We began this chapter by reviewing the research indicating that financial support is strongly correlated with postsecondary completion, a finding that applies to underrepresented minorities as well as others. Data show, however, that in fact the issue of financial support is typically more salient for underrepresented minorities. At a general level, the median household income for underrepresented minorities is lower than for whites and Asian Americans. It can be seen at a more specific level as well in data that illustrate the consequences of insufficient support.

Data from the NSF 2007-2008 Survey of Earned Doctorates, for example, show that in general underrepresented minorities—and African Americans in particular—are more likely to draw on personal and family resources for support when working on a doctorate.

> Differences in the various modes of financial support overall were found among racial/ethnic groups, in part reflecting differences in distributions among broad fields of study (figure 18; table 22). Black doctorate recipients indicated the greatest reliance on their own resources to finance their doctoral program (41 percent), followed by American Indians (32 percent), Hispanics (29 percent), and multiracial recipients (25 percent) (table 22). Asians were the least likely of the racial/ethnic minority groups to report using their own resources (15 percent) (NSF, 2008, 16).[28]

This is true for underrepresented minorities in science and engineering as well. Data on primary mechanism of support for science and engineering doctorates, as shown in Table 5-1, show that underrepresented minorities are twice as likely (20.5 percent) as the average science and engineering doctorate (10.4 percent) to report self-support as their primary mechanism of support. Not surprisingly, as shown in Figure 5-3, underrepresented minorities—and, again, African Americans in particular—report higher debt burdens across fields on completion of a doctorate.

These trends have important consequences. First, self-support and loans create a larger burden for underrepresented minorities both during and after graduation. The need to rely on personal sources—particularly outside work—means that the student has less time to focus on study and research and leads to lower grades, longer time-to-degree, and higher probability of attrition as noncompleters are more likely to have engaged in outside work. Clewell et al. (2005) in their review of the research literature found:

> Studies have shown that holding a part-time job off-campus may be negatively related to persistence in college (Astin 1993), especially for URMs (Nora, Cabrera, Hagedorn, and Pascarella 1996). Pascarella and Terenzini (1991), in their review of the research on this topic, concluded that the evidence suggested that working during college, especially in a job that was related to one's major or career goals, had a positive impact on career choice, attainment, and level of professional responsibility attained early in a career.[29]

---

[28] National Science Foundation. 2008. *Doctorate Recipients from U.S. Universities: Summary Report 2007-08*, p. 16.

[29] B. C. Clewell et al. 2006. *Final Report of the Evaluation of the Louis Stokes Alliances for Minority Participation Program*, p. A-7.

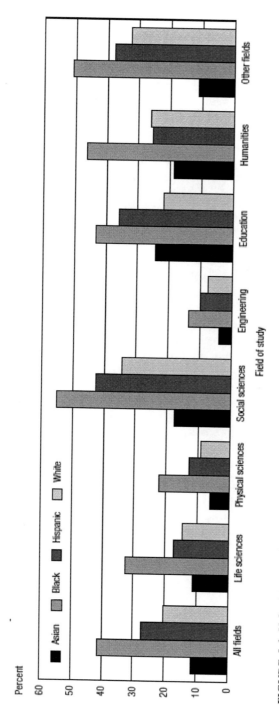

FIGURE 5-3 U.S. citizen and permanent resident doctorate recipients with levels of graduate school debt greater than $30,000, by broad field of study and race/ethnicity, 2008.
SOURCE: National Science Foundation, Doctorate Recipients from U.S. Universities, Summary Report 2007-2008, (NSF 10-309) Arlington, VA: National Science Foundation, December 2009, Figure 19.

Second, and even more important, the burden on the personal finances and debt of those who attend college and graduate school can also serve as a market signal that likely deters other underrepresented minorities from attending, participating, and completing in STEM in the first place, which keeps the proportions of underrepresented minorities in STEM low.

While some of the financial problem can be addressed through need-based programs, there remains a strong need for programs that target underrepresented minorities. Researchers have found that financial incentives are most effective in reducing attrition among low-income and minority students when provided in conjunction with academic support and campus integration, which we will discuss further in the next chapter. The most recent scholarship in this area argues for integrated models of student persistence that recognize the interrelatedness among financial circumstances, academic experiences, student perceptions of their likelihood of program completion, environmental variables, and social support from significant others in the student's family and community. One such model is provided by the NSF's Louis Stokes Alliances for Minority Participation (LSAMP). As shown in Figure 5-4, graduates of LSAMP programs have a higher propensity for additional coursework, graduate enrollment, and graduate degree completion, both in STEM and overall, compared to both white and Asian American students and other underrepresented minority students not in an LSAMP program.

At the doctoral level, again, the package and timing of support are critical. Underrepresented minority students are more likely to receive fellowships than any other type of support and least likely to be supported by research assistantships, as shown in Table 5-2. The use of traineeships and research assistantships, however, can expose more underrepresented minority students to teaching and research experiences and provide opportunities for acquisition of scientific skill, professional development, and social integration into a student's program or department. In general, the availability of a range of financial support options, tailored to the needs of students at a particular point in their graduate studies, can be the most effective way to increase recruitment and reduce attrition of underrepresented minority graduate students in STEM.

Federal programs that provide support to underrepresented minorities in STEM include:

Undergraduate

- *National Science Foundation, Louis Stokes Alliances for Minority Participation (LSAMP)*: This program is aimed at increasing the quality and quantity of students successfully completing science, technology, engineering, and mathematics (STEM) baccalaureate degree programs and

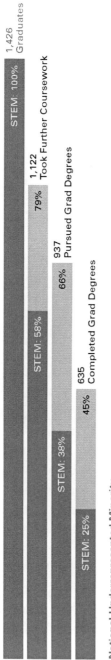

LSAMP Participants

1,426 Graduates — STEM: 100%

1,122 Took Further Coursework — 79% — STEM: 58%

937 Pursued Grad Degrees — 66% — STEM: 38%

635 Completed Grad Degrees — 45% — STEM: 25%

National Underrepresented Minority

36,234 Graduates — STEM: 100%

22,501 Took Further Coursework — 62% — STEM: 43%

16,529 Pursued Grad Degrees — 46% — STEM: 20%

7,139 Completed Grad Degrees — 20% — STEM: 9%

National White and Asian

272,964 Graduates — STEM: 100%

168,145 Took Further Coursework — 62% — STEM: 54%*

120,273 Pursued Grad Degrees — 44% — STEM: 22%

48,315 Completed Grad Degrees — 18% — STEM: 9%

FIGURE 5-4 Graduate coursework, degrees pursued, and degrees completed, LSAMP participants compared to national Underrepresented Minorities and National White and Asian American Graduates.

SOURCE: Clewell et al., *Final Report of the Evaluation of the Louis Stokes Alliances for Minority Participation Program*, Washington, DC: Urban Institute, 2005.

TABLE 5-2 Primary Mechanisms of Support for S&E Doctorate Recipients by Citizenship, Sex, and Race/Ethnicity, 2005

	Research assistantships	Fellowships	Teaching assistantships	Personal	Grant/stipend	Other
Temporary resident	49.1	13.0	17.6	3.1	5.6	11.6
Underrepresented minority, U.S. citizen	12.0	33.8	10.2	19.9	12.6	11.5
Asian, U.S. citizen	24.6	22.0	9.6	11.2	22.7	9.9
White, U.S. citizen	21.8	19.8	13.8	19.7	14.5	10.5
Female, U.S. citizen	16.1	21.9	12.6	23.4	15.2	10.9
Male, U.S. citizen	24.7	21.2	13.5	15.2	14.5	10.9

NOTES: *Personal sources* include personal savings, other personal earnings in graduate school, other family earnings or savings, and loans. *Other* includes employer reimbursement or assistance, foreign support, traineeships, other assistantships, and other and unknown sources. *S&E* includes health fields (i.e. medical and other life sciences). *U.S. citizen total* includes unknown sex. *Underrepresented minority* includes African Americans, Hispanics, American Indians/Alaska Natives, Native Hawaiians/other Pacific Islanders, and multiple races/ethnicities.
SOURCE: National Science Foundation, *Science and Engineering Indicators 2008*.

at increasing the number of students interested in, academically qualified for, and matriculated into programs of graduate study. LSAMP supports sustained and comprehensive approaches that facilitate achievement of the long-term goal of increasing the number of students who earn doctorates in STEM fields, particularly those from populations underrepresented in STEM fields. The program goals are accomplished through the formation of multi-institution alliances. Phase I awards place emphasis on aggregate baccalaureate production. Phase II awards augment the Phase I emphasis with attention to individual student retention and progression to baccalaureate degrees. Phase III awards augment the Phase I and Phase II with attention to aggregate student progression to graduate school entry.[30]

- *National Institutes of Health, Bridges to the Baccalaureate*: The Bridges to the Baccalaureate Program provides support to institutions to help students make transitions at a critical stage in their development as scientists. The program is aimed at helping students make the transition from two-year junior or community colleges to full four-year baccalaureate programs. The program targets students from groups underrepresented in the biomedical and behavioral research enterprise of the nation and/ or populations disproportionately affected by health disparities (targeted groups).[31]

- *National Institutes of Health, MARC Undergraduate Student training in Academic Research (U*STAR)*: MARC U-STAR awards provide support for undergraduate students who are underrepresented in the biomedical and behavioral sciences to improve their preparation for high-caliber graduate training at the PhD level. The program also supports efforts to strengthen the science course curricula, pedagogical skills of faculty, and biomedical research training at institutions with significant enrollments of students from underrepresented groups. Awards are made to colleges and universities that offer the baccalaureate degree. Trainees must be honors students majoring in the biomedical or behavioral sciences who have expressed interest in pursuing postgraduate education leading to the PhD, MD-PhD, or other professional degree combined with a PhD in these fields upon completing their baccalaureate degree.[32]

Graduate

- *National Science Foundation, Alliances for Graduate Education and the Professoriate*: Alliances for Graduate Education and the Profes-

---

[30] http://www.nsf.gov/funding/pgm_summ.jsp?pims_id=5477 (accessed February 25, 2010).

[31] http://www.nigms.nih.gov/Research/Mechanisms/BridgesBaccalaureate.htm (accessed February 25, 2010).

[32] http://www.nigms.nih.gov/Training/MARC/USTARAwards.htm (accessed February 25, 2010).

soriate (AGEP) further the graduate education of underrepresented STEM students through the doctorate level, preparing them for fulfilling opportunities and productive careers as STEM faculty and research professionals. AGEP also supports the transformation of institutional culture to attract and retain STEM doctoral students into the professorate.[33]

- *National Institutes of Health, Bridges to the Doctorate*: The Bridges to the Doctorate Program provides support to institutions to help students make a critical transition in their development as scientists. The program is aimed at helping students make the transition from master's degree programs to PhD programs. The program targets students from groups underrepresented in the biomedical and behavioral research enterprise of the nation and/or populations disproportionately affected by health disparities (targeted groups). The Bridges to the Doctorate Program promotes institutional partnerships between institutions granting a terminal master's degree and institutions that grant PhD degrees in biomedical and behavioral sciences.[34]

- *Ruth L. Kirschstein NRSA Institutional Predoctoral Training Grants*: These graduate programs represent highly diverse areas of basic science and have been judged by peer review to be among the best in the nation. Funds are provided to the institutions, which then administer the training programs. Students apply directly to these programs at the institution and are appointed by the training grant program directors. Trainees receive a base stipend (currently $21,180) that usually is further supplemented by the institution. In addition, each trainee receives an allowance for tuition and fees, health insurance, travel, and training-related expenses.

- *NIGMS Individual Predoctoral Kirschstein NRSA Fellowships to Promote Diversity in Health-Related Research*: These awards, awarded to eligible individual students, support research training leading to the PhD or equivalent research degree, the combined MD-PhD degree, or another formally combined PhD degree. Students must be current matriculants in a biomedically related PhD (or equivalent) program, and strong applicants are those who have already identified their mentor/advisor. The fellowship enhances the diversity of the biomedical, behavioral, health services, and clinical research labor force in the United States by providing opportunities for academic institutions to identify and recruit students from diverse population groups to seek graduate degrees in health-related research. NIH is particularly interested in encouraging the recruitment and retention of the following candidates for this program:

    — Individuals from racial and ethnic groups. Nationally, these
    include, but are not limited to, African Americans, Hispanic Ameri-

---

[33] http://www.nsf.gov/funding/pgm_summ.jsp?pims_id=13646 (accessed February 25, 2010).

[34] http://www.nigms.nih.gov/Research/Mechanisms/BridgesDoctoral.htm (accessed February 25, 2010).

cans, Native Americans, Alaska Natives, and natives of the U.S. Pacific Islands.

— Individuals with disabilities, who are defined as those with a physical or mental impairment that substantially limits one or more major life activities.

A maximum of five years of support is available. NIGMS provides tuition, fees, and up to $4,200 per 12-month period to the predoctoral fellow's sponsoring institution to help defray such trainee expenses as research supplies and equipment.[35]

In addition to these federal programs, foundations have provided important sources of fellowship support for underrepresented minorities in STEM. The Ford Foundation Fellowship Program[36] has been an important source of support at the doctoral level, as has the Alfred P. Sloan Foundation Minority PhD Program.[37] Philanthropy has also been important at the undergraduate level, as has been seen in the Meyerhoff Scholars program, funded by Robert and Jane Meyerhoff and located at the University of Maryland Baltimore County.[38]

While independent evaluations have shown the effectiveness of federal programs such as the NSF Louis Stokes Alliances for Minority Participation (LSAMP) and the NIH minority research training programs, to tackle the scale of change necessary in order to increase underrepresented minority participation in STEM, these and other programs like them must be scaled up to meet the national challenge and achieve the national goal of increasing participation in a transformative way.

---

[35] http://www.nigms.nih.gov/Training/MARC/MARCPredoctoral.htm (accessed February 25, 2010).

[36] http://sites.nationalacademies.org/pga/fordfellowships/ (accessed February 25, 2010).

[37] http://www.nacme.org/sloan/Sloan.aspx?pageid=31 (accessed February 25, 2010).

[38] http://www.umbc.edu/meyerhoff/index.html (accessed February 25, 2010).

6

Academic and Social Support

The NCES study of undergraduate student persistence in STEM discussed in Chapter 2 contrasted the extent to which women and underrepresented minorities major and persist in postsecondary science and engineering programs. It reported different patterns: Although women were less likely to major in STEM fields, they have a slightly higher persistence and graduation rate than that of men; while minorities tend to major at the same rate as nonminorities, their persistence was lower. This ability of women to persist at rates similar to or better than their male peers, NCES observed, was due to similar levels of preparation between males and females before college. With regard to underrepresented minorities, however, a different picture emerges. NCES observed that underrepresented minorities face more barriers to persistence and completion and that postsecondary institutions impact the entire process, from entry to graduation.[1]

Educational attainment is a function of access, information, motivation, affordability, academic preparation and support, social support and integration, and professional development. In their recent book, *Crossing the Finish Line,* Bowen et al. (2009), argue that "educational attainment in the United States is highly consequential. Important are both the overall level of educational attainment and disparities in educational outcomes by race/ethnicity, gender, socioeconomic status (SES), and the kind of university

---

[1] G. Huang et al. 2000. Entry and Persistence of Women and Minorities in College Science and Engineering, National Center for Education Statistics (NCES 2000-601). Washington, DC: U.S. Department of Education.

a student attends."[2] The rate of postsecondary attainment (i.e., receipt of associate's or bachelor's degree) can be increased to meet targets proposed by the College Board, the Lumina Foundation, and President Obama by both enrolling more students in two- and four-year institutions and increasing the percentage of college students who complete. But as Bowen and his colleagues note, we must be concerned not just with overall attainment rates, but with increasing attainment rates across demographic and SES categories: "These outcomes and the forces that drive them are enormously important not only to prospective students and their parents, institutional decision makers, and policy makers but to all who care about both the economic prospects for this country and its social fabric."[3] If we believe in a strong and increasingly important role for science and engineering in developing a strong STEM workforce, educational attainment in these fields, both in general and for underrepresented minorities, is even more important to our future.

ACADEMIC AND SOCIAL INTEGRATION

In *Coming to Our Senses,* the College Board asserts that "colleges and universities have an obligation to improve student retention, minimize dropouts and raise degree completion rates." The report recommends that "what is needed is the development of a culture on campus that includes the expectation that every admitted student will, in fact, graduate, and a determination to understand what is going on when students do not" and argues that "*only the higher education community can address these issues*" (emphasis in original). Further, the report urges a relentless focus "*on the educational needs and challenges of those students most likely to run the risk of dropping out*—low-income, minority or first-generation students. Even after secondary school programs are improved and greater alignment is achieved between K-12 and higher education institutions, it would be foolish to believe that these students, once on campus, will not continue to need additional academic support and advisement."[4] The Education Trust has developed a "seven-step plan" for lowering college dropout rates that was endorsed by the committee. (See Box 6-1.)

These very practical steps to address completion for all students will benefit underrepresented minority students as well; we have seen general efforts that are part of the broader context shape the experiences of underrepresented minority students in STEM.

---

[2] William G. Bowen, Matthew M. Chingos, and Michael S. McPherson. 2009. *Crossing the Finish Line: Completing College at America's Public Universities,* Princeton, NJ: Princeton University Press, p. xiii.

[3] Ibid.

[4] College Board, *Coming to Our Senses,* p. 33.

BOX 6-1
A Seven-Step Plan to Lower College Dropout Rates

No matter their orientation or mission—national research university, regional research university, master's degree institution, or historically black college or university, public or private—different colleges and universities produced substantially different graduation rates, even while enrolling similar students. The Education Trust examined the phenomenon and identified a seven-step process that lowers college dropout rates.

1. Look at your data and act. More higher education decisions should be driven by data. When it is apparent that institutions similar to yours and enrolling similar students are producing different results, it may be time to discard the easy explanations and look for underlying causes on campus. Take student complaints seriously; examine course availability; finish "critical path" analyses that identify "choke points" in curricula and offerings; provide students with online degree audit tools that let them plan degree completion; and make course transfer from elsewhere easier, not harder.

2. Pay attention to details—especially leading indicators. Use technology to track student success. Make course attendance mandatory, track absences, meet with students in trouble, and track data.

3. Take on introductory courses. It's just common sense: If you can get students successfully through year one, their chances of degree completion are much higher. Examine first-year courses. If large numbers or proportions of apparently prepared students are failing, preparation might be the problem, but not necessarily—it could just as easily be a "choke point" of a required course for which not enough sections are provided.

4. Don't hesitate to make demands. Mandatory course attendance is a good idea, as is mandatory lab attendance. At one institution, the faculty, reluctant to require lab participation, found success rates dropped every time the mandatory requirement was waived.

5. Assign clear responsibility for student success. When everyone is responsible, no one is accountable. At one highly successful institution, a central office works with students in challenged high schools and provides summer transition programs and ongoing support and mentoring once enrolled. That office reports to the vice president for student affairs and the vice president for undergraduate education. These students persist to the second year at higher rates than apparently more highly qualified freshmen.

6. Insist that presidents step up to the plate. Institutional leaders have to make sure student success is a priority. Presidents can use the bully pulpit to articulate a vision, insist on data, act strategically and continually "walk the talk." Without presidential leadership (and follow-through on faculty recommendations), efforts to attack dropout rates falter.

7. Bring back the "ones you lose." More common sense—a lot of students who leave without a degree are close to the finish line. The easiest dropout to graduate is the one who is shy of 10 credits or less. One university identified a universe of 3,000 dropouts with at least 98 credits and a GPA of 2.00 or higher.

continued

BOX 6-1 Continued

After tracking down their mailing addresses (relatively easy in the Internet age), the university offered simplified readmission, a degree summary indicating courses required (along with priority enrollment in those courses), and support and counseling. The result: Within a few years, the university could point to 1,800 new alumni and alumnae (including 59 with graduate degrees) and a state impressed with the university's responsiveness.

SOURCE: College Board. 2008a. *Coming to Our Senses: Education and the American Future*, pp. 17-18.

Sustaining Confidence and Self-Efficacy

Within the broader institutional processes of developing a welcoming climate for diversity, institutions, departments, and programs need to focus on how specifically to support underrepresented minority students as aspiring scientists, engineers, and technicians. Over the past several decades, programs have been developed to attract students to STEM majors and provide the necessary support that will enable the students to complete undergraduate STEM degrees and pursue advanced study. Many of these programs have been supported by major federal and private funding agencies, while others have been implemented and supported by individual institutions or departments. In addition to the programs themselves, there is a growing research base on which factors are important elements for broadening participation.[5]

Much of the research has focused on ways to address issues of student motivation and confidence, as the challenges are likely to incorporate psychosocial factors beyond simple questions of access and opportunity. For example, Hurtado et al. (2008) argue that for minority students to become and identify as scientists or engineers, they must negotiate psychological territory that is more complex than it is for majority students. Therefore, interventions that are likely to be successful at broadening the participation of minorities will need to be based upon an understanding of why students choose to pursue certain majors and careers.

Social learning theory explores how individuals acquire social values, recognizing that an individual's personality is based upon unique experi-

---

[5] Chubin, DePass, and Blockus, 2009; Olson and Fagen, 2007. See also http://understandinginterventions.org.

ences, behavior, and cognition (Bandura 1977, 1985; Johnson et al., 1995). Those seeking to influence students' choice of major or career need to recognize the impact of many factors on student choices—including but not limited to formal courses and programs.

One area of focus has been on students' beliefs in their own abilities. This concept, referred to as self-efficacy, has been correlated with issues of persistence and achievement in education settings (Bandura, 1986; Schunk 1981; Zimmerman, 1989; Chemers et al., 2001). Experimental studies in which students were made to enhance their self-efficacy achieved higher performance than those in the control group (Cervone and Peake, 1986; Bouffard-Bouchard, 1990). Thus, one of the key ideas has been to enhance students' confidence in their own abilities. This helps turn the difficulties that students will have to overcome into challenges rather than threats (Chemers et al., 2001).

Both majority and minority students must develop interest in and motivation to pursue science. Then, they must develop the skills to practice science, ably perform science, and, finally, earn the recognition of themselves and others as competent scientists. This is challenging enough. The culture of science on most of our campuses makes this more difficult by constructing a social structure that "weeds out" students in introductory classes and encourages a highly competitive academic atmosphere among undergraduates. Evidence suggests that URM students, under these conditions, experience disproportionate attrition, especially among those who may have been underprepared in high school.[6]

For aspiring minority scientists, academic culture adds several more psychological challenges. First, there is the problem of racial stereotyping. Many teachers and faculty continue to hold low expectations for underrepresented minority students; this can lead to direct barriers to participation, such as when students are excluded from programs, classes, and opportunities. In most cases, these exclusions are not made explicitly on the basis of race, but subtly by not inviting or encouraging students to participate in nonrequired opportunities. But it is not only others who hold these stereotypes; many students internalize these stereotypes about themselves. Thus, different students will view the same situation differently depending on their own background and experiences. In areas such as the STEM disciplines, students may come in with the belief that they will not be able to succeed. This "stereotype threat" can cause students to perform to the level of their internalized stereotype rather than their true abilities. The effect can be especially powerful in situations where students are reminded of the perceived stereotype, even with something as simple as checking a

---

[6] S. Hurtado, N. L. Cabrara, M. H. Lin, L. Arellano, L. L. Espinosa, "Diversifying Science: Underrepresented Student Experiences in Structures Research Programs," *Research in Higher Education* (forthcoming).

box indicating their race or gender prior to taking a standardized test (see, for example, Bonous-Hammarth, 2000; Brown and Day, 2006; Dar-Nimrod and Heine, 2006; Spencer et al., 1999; Steele, 1992; Steele and Aronson, 1995; Steele et al., 2002).

Second, there is the stigma of minority programs. While many minority students welcome the opportunity to participate in programs designed to provide them with opportunities in STEM that they would not otherwise have had or that they need to compensate for earlier poor educational opportunities in STEM, they worry that these programs stigmatize them as somehow less competent than their majority peers who do not require such programs. This is a particular challenge because there is evidence that support from other minorities—including students and faculty members—is one of the most influential factors affecting science ambition and commitment to science (Grandy, 1998).

Third, although in general, underrepresented minorities are likely to find themselves academically and socially isolated, this is more prevalent within STEM (Nettles 1988; Treisman 1992; Cole and Barber 2003). This sense of isolation can result in a lack of a support structure and reinforcement that scientific careers are not for them. Fostering contact with faculty outside of the classroom through both formal mentoring and informal interactions can be helpful in decreasing this isolation. Similarly, building a critical mass of student peers can enhance the social support system as well as student persistence and success (Allen, 1992; Fries-Britt, 2000; Gándara and Maxwell-Jolly, 1999; McHenry, 1997).

Finally, students who come from economically and culturally disadvantaged backgrounds—those who are minorities, are from low-income families, speak English as a second language, or are the first generation in their family to attend college—find themselves in new, often intimidating situations, and often without the same level of information or even access to information that students from advantaged situations take for granted. Even if students are prepared and interested, they and their families may be intimidated by the higher education environment in which they have had little or no previous interaction. This apprehension may, at worst, create barriers to entry or, at a minimum, create barriers to the information needed to be fully successful.

The Tinto Model of Student Retention

Institutions and programs can help to minimize all of these psychological pitfalls to minority participation through initiatives and programs aimed at stimulating student interest and retaining and advancing students in STEM. For example, The Howard Hughes Medical Institute (HHMI) developed a symposia program in which invited participating institutions were asked to provide data on their minority programs. The data collected confirmed that although underrepresented minorities were more likely to drop out of programs early, early intervention strategies made a difference, for example, summer bridge programs, peer mentoring, peer leadership, coaching for social aspects, study groups, early research opportunities, and faculty mentoring.[7]

Clewell et al. described the "Tinto Model of Student Retention," which can be used to provide a theoretical frame for academic and social integration. The LSAMP model utilizes the Tinto model, adapts it to the goal of retaining minority students in STEM majors (by providing supportive, integrative services specific to STEM), and encourages these students to continue on to graduate programs in STEM by providing professionalization opportunities (that is, opportunities to engage in the doing of science as professionals). Clewell et al. continued by describing the role of higher education institutions in encouraging persistence.

The institution can, through its formal and informal structures, assist the social and academic integration of the student and thus encourage persistence in the system. The function of these structures should be to smooth the transition of the student into his or her new environment, encourage the building of learning communities with peers, foster interaction between students and faculty and staff, identify student needs and provide adequate support, and foster academic involvement and learning, among other activities. In outlining his model, Tinto saw the need for retention programs specifically tailored to the needs of different groups of students, such as older students, honor students, students of color, transfer students, and academically at-risk students.[8]

Researchers have modified the Tinto model of student integration and proposed new models to address underrepresented groups and STEM students in particular. For example, Nora et al. (2005) developed the student/institution engagement model as a theoretical framework to examine factors

---

[7] Peter Bruns, Howard Hughes Medical Institute, Presentation to Committee, March 10, 2008.

[8] The authors added:

Much of the research on college student attrition has drawn on the Tinto model, particularly through examining the effects of academic and social integration on students' college persistence or withdrawal. A significant body of studies by various researchers offers support to the validity and usefulness of the theoretical model (Bers and Smith 1991; Braxton, Brier, and Hossler 1988; Cabrera, Castaneda, Nora, and Hengstler 1992; Cabrera, Nora, and Castaneda 1992; Nora, Attinasi, and Matonak 1990; Pascarella, Smart, and Ethington 1986; Pascarella, Terenzini, and Wolfe 1986; Stage 1989; Stoecker, Pascarella, and Wolfe 1988; Williamson and Creamer 1988). Among the few studies in this area that have conducted analyses on minority student populations, Stoecker, Pascarella, and Wolfe (1988) found academic and social integration to be important determinants of persistence, while Nora (1987) found that these factors did not significantly affect retention among Chicano community college students.

impacting withdrawal and persistence decisions of undergraduates past the first year in college.[9] The framework considers precollege factors and pull factors, initial commitments, academic and social experiences, cognitive and noncognitive outcomes, and final commitments as variables.

Institutional Transformation

Efforts to increase minority participation in STEM will have a higher probability of success and produce more robust results in higher educational institutions that incorporate retention strategies, which we will discuss below., However, such institutions have also undergone or are undertaking comprehensive efforts at institutional transformation in their culture by making diversity inclusion a driver within the business functions of the organization. They are creating a welcoming, inclusive environment, inculcating positive attitudes toward and high educational expectations for minority students, and building the capacity for social and educational interaction across racial/ethnic groups that foster success.

Maton et al. (2008) argued further that transformative institutional change is a necessary prerequisite for lasting efforts to affect diversity:

> A subset of theorists have made the case for the necessity of transformative change efforts if enduring progress is to be made in empowering marginalized populations in our society (Hurtado, Dey, Gurin, and Gurin, 2003; Milem and Hakuta, 2000). Maton (2000), for example, has argued that deeply embedded features of social environments influence critical risk and protective processes, nullify person-focused programs, make it difficult to sustain and disseminate promising approaches, and prevent the large-scale mobilization of resources necessary for making a substantial difference. Williams, Berger, and McClendon (2005) argue that a series of transformations are required in organizational culture and behavior if campus diversity initiatives are to make a difference; otherwise, possible benefits of such initiatives may fade very easily. Ibarra (2001) makes the case that only a fundamental change in the culture of higher education related to diversity will result in substance advances for minority students.[10]

The authors describe how ongoing dialogue within a campus community on issues related to race, a strengths-based rather than a deficits-based

---

[9] A. Nora, L. Barlow, and G. Crisp. 2005. Student Persistence and Degree Attainment Beyond the First Year in College: The Need for Research. In A. Siedman's (Ed.), *College Student Retention: Formula for Student Success*. Praeger Publishers, pp.129-154.

[10] K. L. Maton, F. A. Hrabowski, M. Ozdemir, and H. Wimms. 2008. Enhancing Representation, Retention, and Achievement of Minority Students in Higher Education: A Social transformation Theory of Change, In M. Shinn, & H. Yoshikawa, H. (Eds.), *Toward Positive Youth Development: Transforming Schools and Community Programs*. New York: Oxford University Press, pp. 115-132.

view of minority students, and intensive data-based reviews of minority student achievement are all useful in implementing transformative institutional change.

Hurtado et al. (1999) identify four key steps institutions must take to promote an improved campus climate for diversity:[11]

1.   "Affirm the goal of achieving a campus climate that supports diversity as an institutional priority." A campus-wide commitment to inclusiveness provides the best environment for planting the seeds of diversity. This should be articulated by university leaders—faculty, department chairs, deans, provosts, chancellors and presidents, and governing boards (trustees and regents)—both in the university mission and in every day affairs. The visible and continuing commitment of campus leaders to diversity and to minority participation provides the overall, critical tone that signals appropriate actions for others. Faculty are important in the production of diversity in the student population—particularly at the PhD level—as they determine who will be the next generation of scientists and engineers. There can be a large disconnect between what leaders say and what faculty do, and the direct connection, with faculty buy-in, must be made.

2.   "Engage in a deliberate, self-conscious process of self-appraisal that will provide a baseline of information on the current state of affairs regarding the campus climate for diversity," with a focus on both underrepresented minorities and women.[12]

3.   "Guided by research, experiences at peer institutions, and results from the systematic assessment of the campus climate for diversity, develop a plan for implementing constructive change that includes specific goals, timetables, and pragmatic activities." Such activities could include the development, implementation, and enforcement of admissions policies that reinforce diversity within the legal parameters of the Michigan decisions in order to ensure a significant and sufficient overall level of minority participation on campus, and rewarding faculty in the promotion and tenure process for developing student talent, both in general, and for underrepresented groups, including minorities; and providing support and retention measures for underrepresented minority students.

---

[11] Hurtado et al. 1999. *Enacting Diverse Learning Environments: Improving the Climate for racial/ethnic Diversity in Higher Education*, ASHE-ERIC Higher Education Report Volume 26, No. 8, Washington, DC: The George Washington University, Graduate School of Education and Human Development.

[12] In *Beyond Bias and Barriers* (2007), the National Academies recommended that institutions implement self-assessments for evaluating how well they are serving women and minorities in science and engineering.

4.   "Implement a detailed and ongoing evaluation program to monitor the effectiveness of and build support for programmatic activities aimed at improving the campus climate for diversity."

Chubin and Malcom (2008) present three issues that institutions must address to achieve better representation of minorities in STEM:[13]

1.   The educational case for diversity, showing how students and society benefit from it. The institution can then determine a strategy. What policies should be altered, what practices endorsed, what structural changes made, and what resources committed.
2.   Thinking holistically about diversity in STEM, including the need for everyone on our campuses to be exposed to diverse ideas and worldviews. Functions such as admissions, financial aid, and faculty recruitment and advancement should be reexamined and share responsibility for that goal.
3.   Acknowledging that stereotypes still matter and affect perceptions of quality and expectations for performance.

Efforts to promote inclusivity, however, are not enough unless they are carried out through proactive efforts to encourage the social interaction that is needed to realize inclusivity and the benefits to students of peer-to-peer and faculty-student interactions. Peer-to-peer interaction can help increase cross-racial understanding, reduce barriers to integration in educational and extracurricular activities, and improve retention and success. Faculty-student interaction promotes the development of educational aspirations, academic achievement, persistence, and self-concept.

Thus, to quote Hurtado et al. (1999) further, institutions should [14]

1.   Involve faculty in efforts to increase diversity that are consistent with their roles as educators and researchers.
2.   Increase students' interaction with faculty outside class by incorporating students in research and teaching activities.
3.   Create a student-centered orientation among faculty and staff.
4.   Initiate curricular and co-curricular activities that increase dialogue and build bridges across communities of difference.
5.   Include diverse students in activities to increase students' involvement in campus life.
6.   Increase sensitivity and training of staff who are likely to work with diverse student populations.

[13] D. E. Chubin and S. Malcom. 2008. Making a Case for Diversity in STEM Fields. http://www.insidehighered.com/layout/set/print/viwwews/2008/10/06/chubin.
[14] Hurtado, et al.

BOX 6-2
Broadening Participation in Graduate School

Recommendations for Institutions of Higher Education:

- Closely monitoring completion and attrition rates of students from underrepresented groups and implementing best practices to improve completion rates
- Developing training programs for graduate student mentors who can help a diverse group of students navigate graduate school successfully
- Experimenting with programs that use technology, which attracts and appeals to today's students
- Identifying strategies for recruiting a more diverse faculty by broadening faculty search criteria and by advertising positions as widely as possible
- Identifying possible faculty members by establishing linkages with specialized targeted institutions, including HBCUs
- Encouraging faculty to be ever vigilant of opportunities to promote a more inclusive environment for students as well as themselves
- Encouraging graduate deans who are uniquely positioned in institutions of higher education to become leaders in inclusiveness by:
  - Working to ensure that inclusiveness is a team effort in the institution, involving the student body, faculty, and the highest levels of the administration
  - Supporting the development of a more inclusive curriculum with courses that appeal to a wide range of students
  - Using their understanding of the academic pipeline to assist in diversifying the faculty
- Continuing to foster partnerships with those in the business community who have made inclusiveness an essential part of their organizations
- Continuing to develop strategies that are effective in helping to make graduate education responsive to the intellectual aspirations of all students
- Recognizing that broadening participation is a dynamic process and that supporting diversity and inclusiveness is a priority. In this increasingly global community, developing culturally competent graduates, faculty, and administrators is integral to continued U.S. leadership.

SOURCE: Council of Graduate Schools. 2009. *Broadening Participation in Graduate Education.*

As shown in greater detail in Box 6-2, the Council of Graduate Schools has provided additional recommendations for increasing diversity in graduate programs. Evidence of the cultural transformation that results from these efforts can be seen most readily in observable statistics regarding minority enrollment, graduation rates, faculty hiring, and the like. These are key both as indicators of progress and signals to the larger community of commitment and change.

CONCLUSION

7

The Journey Beyond the Crossroads

A strong and robust science and engineering workforce drives the nation's ability to thrive in a competitive, knowledge-driven global economy. As demonstrated in previous chapters, the nation needs to pursue aggressive strategies to ensure greater participation of underrepresented minorities in that STEM workforce and to equip them with the technical competencies for emerging needs.

We therefore suggest a need to realign national policies and practices and to integrate these policies and practices vertically and horizontally. The logic to accomplish this feat includes principles to guide the development of transformative programs and activities, description of institutional roles as enablers in the production of minorities in STEM, and characteristics of programs that are designed for optimal impact.

PRINCIPLES

1. *The problem is* **urgent** *and will continue to be for the foreseeable future.* To be proactive in shaping our future requires that we make broadening participation a national priority. The demographics alone signal immediacy. Acting *now* to affect the pathways of *today's* elementary school students will change the educational outcomes of high school graduates in 2020. Acting *now* to improve the educational pathways of *today's* high school students will impact the doctoral class of 2020. Given the long time horizon for demonstrable results of efforts to improve the participation and success of underrepresented minorities in STEM, we cannot delay if

we want to get ahead of the workforce challenges and opportunities that are coming in the next decade.

2. *A successful national effort to address underrepresented minority participation and success in STEM will be sustained.* We worry that after an initial effort to address underrepresented minority participation in STEM, national attention may turn to some other crisis of the day and that initial momentum as well as incremental gains may be lost. In its landmark 2003 case, *Grutter v. Bollinger,* the Supreme Court wrote: "The Court expects that 25 years from now, the use of racial preferences will no longer be necessary to further the interest approved today." The year 2028 is still almost two decades away; until that day when we will no longer need to focus on the participation of underrepresented minorities to ensure strength and equity in our science and engineering workforce, a deliberate national effort is needed to galvanize stakeholders and resources toward this end.

3. *The potential for losing students along the pathway from preschool to graduate school necessitates a comprehensive national approach focusing on all segments of the pathway, all stakeholders, and the potential of all programs, targeted or nontargeted.* Understanding that race and ethnicity—and all that group identity may mean for social, economic, and educational opportunity—comprise a key dimension of STEM educational attainment provides an important point of leverage for considering STEM education policy. Indeed, focusing on underrepresented minorities as a point of leverage in STEM education policy allows us to revisit existing education programs from a new perspective. As shown in Table 7-1, there are four existing approaches to the issue. In the first quadrant are policies that seek to affect education across fields for all groups. In quadrant two are policies and programs that seek to improve the educational opportunities across fields, but in particular for underrepresented minorities. In the third quadrant are policies and programs designed to improve science and engineering education for all groups. In the fourth quadrant are policies and programs specifically targeting underrepresented minorities in science and engineering.

Federal and state education policies and programs that affect underrepresented minorities, including those in STEM, can be identified in each of these quadrants. For example:

1) All Fields/All Groups: Universal Preschool, No Child Left Behind Act, Pell Grants

2) All Fields/Underrepresented Minorities: Affirmative Action, Top 10 Percent Admissions Rule (e.g., California and Texas policies for UC and UT undergraduate admissions)

TABLE 7-1 Approaches to Increasing Underrepresented Minority Participation and Success in Science and Engineering

		Demographic Target	
		All Groups	Underrepresented Minorities
Fields	All fields	1	2
	Science and Engineering	3	4

3) STEM Fields/All Groups: SMART Grants, NSF Integrative Graduate Education and Research Traineeship Program, NIH National Research Service Award Graduate Fellowships

4) STEM Fields/Underrepresented Minorities: NSF Louis Stokes Alliance for Minority Participation (LSAMP), Alliance for Graduate Education and the Professoriate (AGEP), HBCU-UP, TCU-UP, NIH Minority Access to Research Careers

The nation can utilize all of these programs and their methodologies more synergistically to accomplish the goal of broadening participation, using targeted programs as necessary but also embedding the goal of increased participation in nontargeted programs. In particular, all of the nation's higher education institutions can make underrepresented minority participation and success in STEM a priority and take actions necessary to become more inclusive.

4. *Students who have not had the same degree of exposure to STEM and to postsecondary education require more intensive efforts at each level to provide adequate preparation, financial support, mentoring, social integration, and professional development.* Effective policies, strategies, and interventions are needed to target each segment of the STEM education pipeline trajectory, from preschool to graduate school. They must aim to reverse the downward spiral in academic achievement for the nation in general, but particularly for underrepresented minorities. In addition, they must target the perpetual problems of elementary school readiness, achievement gaps, college preparedness, cultural diversity, and attrition and degree completion in STEM. Ingredients for success in STEM education are discussed in detail in Appendix F.

Although there is still much to understand about how students learn and how to improve retention and completion in educational programs, we already do know a lot about what works. In fact, many interventions that assist students from one background are the same as those that would assist students from any background: It helps to be prepared, informed, and motivated; to have financial, social, and academic support; and to have institutional resources necessary for success once enrolled in the field. Yet, there are issues that are specific to STEM, for example, how to teach science and mathematics so that students learn and sustain interest. And there are issues that are specific to *underrepresented minorities* who have not had the same degree of exposure to STEM and to the world of postsecondary education, or who, for whatever reason, may feel, or be made to feel, like an "outsider" and require more intense efforts at each level.

Effective interventions for minorities in STEM are well documented in reports such as that of Chubin, DePass, and Blockus (2009), presented at an AAAS conference.[1] This report is a compendium of research reports and articles from an interdisciplinary community of scholars and was designed not only to inform practice and research but also to inform practice with research. The report demonstrates the importance and opportunity for stakeholders across all segments to contribute to this national effort.

5. *A coordinated approach to existing federal STEM programs can leverage resources while supporting programs tailored to the specific missions, histories, cultures, student populations, and geographic locations of institutions with demonstrated success in the preparation and advancement of underrepresented minorities in STEM.* The most recent inventory of federal STEM education programs, developed by the Academic Competitiveness Council (ACC), catalogued 105 such programs across 12 agencies with an aggregate funding of $3.12 billion for fiscal year 2006. By educational level, this inventory includes:

- Kindergarten through grade 12:   24 programs
- Undergraduate and graduate education: 70 programs
- Informal education: 11 programs

Slightly more than half of the inventory, 57 programs, either target underrepresented groups (underrepresented minorities, women, or persons with disabilities) or include increasing the participation of these groups in STEM as an embedded goal.[2]

---

[1] D. Chubin, A. L. DePass, and L. Blockus. 2009. *Understanding Interventions That Broaden Participation in Research Careers.* Volume III. Summary of a Conference, Bethesda, MD, American Association for the Advancement of Science.

[2] The ACC inventory appears to have missed several programs at the National Institutes of Health, so the overall list is likely longer, as is the list of programs with a focus on underrepresented groups.

The ACC (2007) found that "many of these programs share similar goals" and that "while duplication is not inherently bad . . . coordination among agencies could be improved."[3] The Council identified two reasons why coordination would be helpful, and our own experiences with STEM education also suggest a third:

- Grants often support projects that appear uninformed by similar earlier experiences.
- Agencies with similar STEM programs and goals sometimes do not share information about the work they fund.
- Agencies with similar missions sometimes fund programs on the same campus, even targeting the same population, without any coordination of activities.

Institutions and students would be better served by operational coordination of STEM education programs among agencies, including joint funding competitions. This coordination could provide for effective articulation of programs that target different educational stages and reduce redundancy, increase effectiveness, and allow for leveraging of funds as appropriate.

Coordination of STEM education programs, including those focused on increasing the participation and success of underrepresented minorities, can be accomplished in several ways:

1. A Committee on STEM Education within the NSTC can help agencies share information on effective practices and also develop partnerships that leverage resources and increase impact.

2. Bilateral partnerships between agencies can elevate the stature and catalyze national momentum of STEM initiatives. A Memorandum of Understanding between the NSF and NASA that provides for cooperation and coordination of STEM education programs is an example of this type of practice.

3. Within an agency, partnerships can also advance STEM education goals. For example, the NSF's Integrative Graduate Education and Research Training (IGERT) program is administered as a cross-directorate program out of the Education and Human Resources (EHR) Directorate. Similarly, the Geosciences Directorate's GEO Education and Diversity Strategic Plan (2010-2015) is aligned with investments being made within EHR and proposes to partner with agencies such as NASA, NOAA, DOE, and USGS. Partnerships across other directorates can be developed to leverage the agency's resources to optimize the broadening participation strategy.

---

[3] Academic Competitiveness Council, Report of the Academic Competitiveness Council, Washington, DC: U.S. Department of Education, May 2007. http://www2.ed.gov/about/inits/ed/competitiveness/acc-mathscience/report.pdf (accessed February 19, 2010), p. 3.

4. Finally, one or more agencies can provide funding to an institution or group of institutions to better integrate activities focused on engaging minorities in STEM. Although not specifically directed to broadening participation, the NSF's Innovation through Institutional Integration (I^3) initiative illustrates this strategy.

While greater coordination and strategic partnerships can make both national and local efforts more effective and powerful, these efforts must be well conceived, leveraging programmatic strengths while retaining the intrinsic power found in the focus of individual programs designed to meet specific needs. Thus, it would be a mistake to consolidate programs, tailored to the specific missions, histories, cultures, student populations, and geographic locations of HBCUs, TCUs, and HSIs that have demonstrated success in the preparation and advancement of groups underrepresented in STEM.

6. *Evaluation of STEM programs and increased research on the many dimensions of underrepresented minorities' experience in STEM help ensure that programs are well informed, well designed, and successful.* Federal agencies, higher education institutions, professional associations, and philanthropy can drive efforts to increase the participation of underrepresented minority students in STEM through program evaluation, identification of best practice, information dissemination activities, and support for inquiry that focuses on key areas of research.

Program evaluation (summative and formative) is a useful tool for both policy making and program management. Evaluation can:

- Provide real-time feedback on program design, processes, and implementation;
- Assess whether a program and particular program features are successful or not; and
- Provide information that is useful to a program and that can be shared with others with similar programs or those who desire to develop one.

The ACC has argued that there is significant room for increased and more rigorous evaluation of federal STEM education programs in general. Building Engineering and Science Talent (BEST) (2005) found that there also has been little such evaluation of programs to increase the participation of underrepresented minorities in STEM.[4]

---

[4] Building Engineering and Science Talent, A Bridge for All: Higher Education Design Principles for Broadening Participation in Science, Technology, Engineering, and Mathematics. San Diego, CA: February 2004. http://www.bestworkforce.org/PDFdocs/BEST_BridgeforAll_HighEdFINAL.pdf (accessed December 22, 2009).

Indeed, the number of rigorous evaluations of programs designed to increase the participation of underrepresented minorities in STEM is small, even including three large efforts undertaken since the publication of *A Bridge for All*, an *Assessment of NIH Minority Research and Training Programs* by the National Research Council, and evaluations by the Urban Institute of the Louis Stokes Alliances for Minority Participation (LSAMP) and Historically Black College and University Undergraduate Program (HBCU-UP) programs at the NSF.[5] Evaluations of similar efforts by private foundations are also warranted to explore opportunities to find partners in funding the most promising programs.

A corollary to the importance of program evaluation is the dissemination of information about practice that is derived from these evaluations and from other research. The development and maintenance of a database or clearinghouse of information from evaluation and research could enhance the accessibility of evidence-based approaches to formulating programs and strategies and could by extension significantly enhance effectiveness.

Further research into the many dimensions of the experience of underrepresented minorities in STEM also will inform policy and practice in positive ways. The report already has drawn on a growing body of research on the social, cultural, psychological, economic, and educational dimensions of increasing participation and success. We have presented selected researchers and scholars in Appendix H and outlined priority areas of inquiry for future research. These include mentoring, social support networks, institutional and departmental culture, attrition, and the characteristics of minority-serving institutions that enable them to nurture and sustain underrepresented minorities in STEM. Suggestions also include the need for additional research on the interrelationship of gender and race/ethnicity in STEM, developing a critical mass of underrepresented minority students in a program, and the impact of intervention programs. Further research into the contributions of eminent underrepresented minority scientists and engineers and how their examples and experiences affected other minorities in STEM would provide additional useful insights.

INSTITUTIONAL ROLES

The committee was charged with discussing the role of minority-serving institutions (MSIs) in increasing underrepresented minority participation and success in STEM. To do that, we must discuss MSIs in the context of

---

[5] National Research Council. 2005. *Assessment of NIH Minority Research Programs; Phase 3*. Washington, DC: The National Academies Press. Beatriz Chu Clewell et al. 2006. *Revitalizing the Nation's Talent Pool in STEM* . Urban Institute: Washington, DC. Beatriz Chu Clewell et al. 2010. *Capacity Building to Diversify STEM: Realizing Potential Among HBCUs*, Urban Institute: Washington, DC.

all higher education institutions. We would like to share four observations at the outset regarding their respective roles.

1. An analysis of the baccalaureate origins of underrepresented minority PhDs in STEM (included in Appendix G) finds that institutions successful in graduating such students at the baccalaureate level are diverse. For African American and Hispanic S&E PhDs, top baccalaureate origin institutions included both minority-serving and predominantly white institutions. For example, as shown in Table 7-2, about one-third of the baccalaureate institutions for African American PhDs in STEM fields were HBCUs, and about two-thirds were non-HBCUs.

An NSF analysis that normalized baccalaureate origin rankings by percentage of bachelor's degrees awarded to African Americans also showed that among PWIs, both research universities and liberal arts colleges contributed to the undergraduate education of future doctorates, as was the case for HBCUs.

The analysis also shows that those institutions—whether minority-serving or predominantly white—that are successful are doing something special. What they are doing is not a mystery, as will be discussed below, and can be replicated at other institutions. Research on underrepresented minority students in the STEM pathway indicates that although these students enroll at rates similar to those of their Asian American and white counterparts, they drop out at a much higher rate. The major contribution of the top baccalaureate producers of PhDs, then, lies in their ability to *retain* underrepresented minority undergraduates in the natural sciences and engineering. The analysis in Appendix G shows that they appear to do it through a focus on STEM education in one or more particular fields that represent the core strengths of the institution.

2. The challenge of increasing underrepresented minority participation and success in STEM is so substantial that it requires every institution to step up to the plate regardless of its size or type. This is a responsibility for the nation, and every institution must be held accountable.

3. Numbers are not enough. It is equally important, if not more so, that we ultimately focus on quality. Our aim must be to produce underrepresented minority students from all types of institutions at bachelor's, master's, and doctoral levels who are strongly qualified for the STEM workforce, advanced training, and research. The support of an advocate mentor while pursuing opportunities in the job market—particularly the academic job market—can be invaluable for underrepresented minorities.

The diversity of American higher education institutions is a competitive advantage in the global knowledge economy. This institutional diversity could be, but is not yet, effective in addressing the varied needs of under-

TABLE 7-2 Number of U.S. Baccalaureate Institutions of African American PhDs in Science and Engineering, by Broad Field and Institutional Type, 2006

Institutional Type	Social & Behavioral Sciences	Natural Sciences & Engineering	Science and Engineering
HBCUs	93	162	255
Non-HBCUs	257	267	524
Total	350	429	779
% HBCU	26.6%	37.8%	32.7%

Note: Totals and percentages do not include unknown institutions.
Source: NSF/SRS, WebCASPAR.

represented minority students. Currently, only a small number of institutions are playing their potential roles. Everyone else is failing underrepresented minorities, their institutions, and America.

Predominantly White Institutions

We need to increase *retention* of African American, Hispanic, and Native American students in NS&E fields on a large scale to influence their numbers in science and engineering, particularly at the doctoral level. The best way to do so is to replicate programs, resources, and focused efforts at the successful PWIs at *a very large number* of similar institutions, especially large state flagships (which could produce larger numbers and be more economical for students to attend). They also can learn from the MSIs that have proven success in producing large numbers of minority students in STEM. As outstanding as individual institutional strategies are at institutions such as UMBC, Georgia Tech, Rice, and MIT, they individually contribute only marginal change to a huge problem. What is needed is for *every* four-year institution to develop and implement its own version of programs with demonstrated and sustained success such as the UMBC Meyerhoff, Georgia Tech Focus, or Rice University Computational and Applied Mathematics (CAAM) programs. (See Box 7-1 for a detailed description of the CAAM program.) Each of these had a single or initial champion that helped to drive the numbers that these institutions have been able to produce.

Institutions can use the program guidance described later in this chapter to develop effective interventions, perhaps focusing on a specific field of science or engineering that is a special strength of the institution. Majority schools can enable minority students by providing them with the quality educational experiences that they provide to majority and international students. They can act affirmatively, removing systemic barriers to the partici-

BOX 7-1
Rice University
Computational and Applied Mathematics Program

The American Mathematical Society (AMS) presented Rice University's Computational and Applied Mathematics (CAAM) Department its 2010 "Mathematics Programs That Make a Difference" award in acknowledgment of "the department's unwavering commitment to students through individual guidance and support" that "has created an exceptionally welcoming community in which students thrive." For 25 years, the Rice University Computational and Applied Mathematics Department has worked to increase participation of underrepresented minority (URM) students at the PhD level. Over those 25 years, 34 URM PhDs (6 African American, 15 domestic Hispanics, and 13 Latin American Hispanics) have been produced. An additional 33 women have received CAAM PhDs over this period.

CAAM URM graduates have distinguished themselves across the country in government labs, industry, and university faculties, many in positions of leadership. Also, the CAAM department program has served as a model for first a university-wide, then a Houston-wide program across all science, technology, engineering, and mathematics (STEM) disciplines, and for an engineering-wide program at the University of Wisconsin-Madison (UW-M).

Program Vision: Admitting a full spectrum of underrepresented minority students, some of which would be rejected using traditional admissions criteria, and then creating a community that provides academic, social, and personal support are the cornerstones of the CAAM program. The goal was to find the "diamonds in the rough" so as to increase participation nationally, not just to compete with other good schools for the few stellar students that would be accepted at any elite school in the country.

Admissions: CAAM admissions decisions are made by the CAAM Graduate Admissions Committee, with input from a central committee that is part of Rice's Alliances for Graduate Education and the Professoriate (AGEP) program, that advises on minority graduate admissions across all science and engineering departments. The AGEP committee controls approximately 16 graduate minority fellowships and tuition waivers funded by Rice each year, and the CAAM department, as well as all other STEM departments, sends applications to this committee for consideration. Since department graduate admissions committees tend to be rotating, and new committee members may not understand or share the goal of diversity, this standing committee provides continuity of purpose and understanding on diversity matters.

The committee takes a holistic approach to evaluating students for admissions; standardized test scores, undergraduate grades, quality of undergraduate institution, and letters of recommendation are all reviewed as a whole. For GRE scores, the committee chooses a threshold score at which students should be successful. Students with scores significantly above the threshold are deemed to be equivalent, relative to the test score; the score is dismissed, and admission decisions are guided by the other criteria. Students with scores near the threshold value are considered with extra care, and students with scores significantly below the threshold, have to have very strong credentials otherwise to be accepted.

Admissions is still an art rather than a science, but experience and better understanding of how to evaluate is gained with each new class. Having the input of a knowledgeable and caring minority committee lessens the mistakes of excluding people with the ability to succeed or admitting people who do not. Having the strength of funding behind them gives the AGEP committee some clout with the department about decisions. Current underrepresented minority students also play a major role in the recruitment of new students. They recruit at national meetings in coordination with departmental recruiters. Moreover, they play a role in hosting and entertaining visiting underrepresented minority students who have been accepted by the various departments.

Retention: No quality captures the essence of the CAAM program like that of *community.* Incoming URM students are brought to Rice during the summer prior to their first year. That summer is spent working on a research project, but the primary purpose is to help students develop a support system before they start classes. Students who are more senior mentor the incoming students in this acclimatization. Bringing together students and faculty from all STEM departments creates a critical mass for community, and concerns among the graduate students across STEM disciplines are often common enough to share as support for one another. These weekly sessions include guest speakers, student research presentations, social interactions, and professional development activities.

Faculty Involvement: Strong faculty involvement is a key component of the Rice CAAM model. The CAAM program creates close student-faculty relationships early in students' careers with minority and other caring faculty to build trust for any future interventions. What has proven successful at CAAM is that the faculty program leaders keep close watch on students and proactively check on their progress. They then make recommendations such as study groups, tutoring, a reduced course load, and undergraduate courses, even changing research advisors or stepping in to co-advise. Important in all of this is that no stigma accompanies these recommendations. Students frequently emerge from these actions strong and on par with other CAAM students.

— Richard Tapia, Rice University

pation of underrepresented minorities in college by developing admissions and financial aid policies that promote diversity in the campus population. They can also provide the same kind of warm, supportive environment for underrepresented minority students that these students find at minority-serving institutions.

Liberal arts colleges can provide STEM programs for students who want to pursue a science-based career or prepare for professional programs or graduate school in science or engineering. Master's-focused institutions and research institutions can provide professional science master's programs that offer a graduate-level opportunity for students interested in science who want to work in industry, government, or nonprofits. Master's-focused institutions can also provide a bridge to doctoral study for minority students by providing them with a solid introduction to graduate school in an environment that prepares them for success later at the doctorate level.

To address the particular needs of underrepresented minority students, majority institutions can adopt more specific guidelines that include:

- Senior administrators, especially science and engineering deans, actively endorsing and supporting minority-focused programs in order to promote faculty buy-in.
- Respected faculty members in the STEM fields acting as mentors, advisors, role models, and advocates, along with a policy of insisting that faculty and others hold everyone to the same high standards and expectations that encourage all students to perform at their best level (Tapia, 2009).[6]
- Support systems that enable students to fully assimilate into the culture of the institution and mainstream into campus life.

Minority-Serving Institutions

Minority-Serving Institutions (MSIs) as defined by 20 USC Sec. 1067k[7] play a distinctive role for underrepresented minorities in STEM. (See Box 7-2 for definitions of MSIs.) They have a legacy of recruiting, retaining, and graduating a disproportionate number of minorities, especially at the undergraduate level. MSIs enroll approximately 60 percent of all minority undergraduates at two-year and four-year institutions[8] and are the baccalaureate origin of a large segment of minority STEM doctorate

---

[6] Richard Tapia, "Minority Students and Research Universities: How to Overcome the 'Mismatch," *The Chronicle of Higher Education*, March 27, 2009.

[7] 20 USC Sec. 1067k.

[8] National Center for Education Statistics. 2008. *Characteristics of Minority-Serving Institutions and Minority Undergraduates Enrolled in These Institutions*, Washington, DC: U.S. Department of Education.

BOX 7-2
Minority-Serving Institutions

Minority Serving Institutions are themselves diverse, and policies and programs aimed at serving underrepresented groups should leverage these differences.

- The **Tribal Colleges and Universities (TCUs),** for example, were established to respond to the needs of the American Indian population in geographically isolated communities based primarily on reservations. "TCUs have an additional mission: They serve as a venue for educational attainment for American Indian students and are committed to the preservation and resuscitation of native cultures and traditions."[a] However, they are plagued by low college access and degree completion rates, inadequate financial support, and historical discrimination. They rely on federal intervention because they typically are located on federal trust territories and cannot access state funds or local tax levies.

- **Historically Black Colleges and Universities (HBCUs)** typically date from the nineteenth century and were established to serve African Americans who were excluded from white institutions. These institutions are predominantly though not entirely in the South, stretching from Pennsylvania to Texas. Their student population is overwhelmingly African American. Yet they, too, are diverse, ranging from very small institutions to research universities like Howard and Florida A&M that operate doctoral and professional programs.

- **Hispanic-Serving Institutions (HSIs)** differ definitionally from TCUs and HBCUs in that, in order to meet the federal definition, an institution typically has enrollment of Hispanics at or above 25 percent of the student population. HSIs can, therefore, be both predominantly white and Hispanic-serving at the same time. This, of course, creates a different dynamic. As with HBCUs, HSIs also are diverse, ranging from those with relatively small Hispanic population to those that are overwhelmingly Hispanic (e.g., University of Puerto Rico) and from those that are relatively small to much larger doctoral institutions, such as the University of Texas at El Paso.

---

[a]The Path of Many Journeys: The Benefit of Higher Education for Native People and Communities (2007) Institute for Higher Education Policy.

---

recipients, despite the fact that many are smaller than mainstream institutions and receive significantly fewer federal obligations for R&D and science and engineering. Much of their success is attributed to their mission to educate underrepresented minority students while providing a sensitive climate, role models, and emphases on teaching, peer support, mentoring, and service to their communities.

In spite of their importance to higher education and to society, many MSIs are struggling financially, often lacking adequate resources to enable them to offer competitive salaries to their faculty and staff, provide scholarships and fellowships to students, and maintain robust research infrastructures. With additional support, they can expand their effectiveness in recruiting, retaining, and graduating an increased number of minorities in STEM, especially at the baccalaureate level. Partnerships among MSIs and between MSIs and PWIs can be effective in building a sustained pipeline of minorities in STEM when these partnerships provide increased access to research opportunities for faculty and students, faculty and student exchanges, and knowledge transfer.[9] Increasing the proportion of faculty at MSIs who are themselves underrepresented minorities can provide both greater opportunity and new role models for aspiring underrepresented minorities in STEM.

Historically Black Colleges and Universities

The fact that HBCUs enroll smaller percentages of African American students in S&E majors than do PWIs but graduate a larger percentage speaks to the efficacy of these institutions in retaining these students. HBCUs by their very mission, purpose, and environment are more likely to achieve success because the nurturing, individualized nature of instruction, and the presence of a critical mass of African American students and role models among faculty may offset the lack of resources. A recent evaluation of NSF's HBCU-UP program has shown that providing resources to strengthen the institutional STEM infrastructure at these institutions has resulted in their producing students who enter STEM graduate programs at a greater rate than do institutions of higher education nationally. (A similar phenomenon exists with women's colleges, which are more likely to produce women who ultimately attain doctorates in the sciences.) UMBC, Georgia Tech, and MIT may incorporate some of the attributes of HBCUs, but they have adapted their strategies to be successful in a PWI environment. Other PWIs should learn from this success.

The HBCU distribution across Carnegie classifications suggests that a strategy for utilizing this resource for increasing underrepresented minority success in STEM should be sensitive to differences within this group and build on the strength of institutions best positioned to advance students in these areas. A small number of HBCUs are particularly successful in the baccalaureate production of African American STEM doctorates. (See Box 7-3 for a description of Florida A&M's student support pro-

---

[9] National Research Council. 2009. *Partnerships for Emerging Research Institutions*. Washington, DC: The National Academies Press.

BOX 7-3
Life-Gets-Better at Florida A&M University

The Presidential Scholars Program at Florida A&M University is a student-centered program of academic scholarships for student enhancement and empowerment. This program offers a range of seven unique scholarships and seeks to identify and award scholarships to students who have excelled at the national achievement level. It also seeks those committed students who have achieved a measure of success and have the potential to pursue a rigorous program of study, particularly in science, technology, engineering, and mathematics (STEM).

The premier scholarship in the Presidential Scholars Program is the Life-Gets-Better Scholarship, which is offered to students seeking a STEM major. The Life-Gets-Better Scholarship is a unique award at Florida A&M University. It provides a full scholarship and summer internship to a student who has outstanding academic achievements and is jointly sponsored by Florida A&M University and major corporations. The sponsoring corporations provide the students with paid summer internships. Each summer, the student continues employment with the same firm at an increasing rate of pay. The four-year scholarship covers the cost of tuition and fees, room and board, $500 per semester stipend, and a book scholarship each semester.

The Presidential Scholars Program also has scholarships for students who have demonstrated academic potential and have achieved success. Each of the seven scholarships in the Presidential Scholars Program has elements of the Life-Gets-Better Scholarship. The nurturing environment that is imbedded in the Presidential Scholars Program at Florida A&M University emphasizes twenty-first century skills required for knowledge-based STEM careers and readiness for graduate school and for lifelong learning, while assuring the acquisition of professional attributes required for industry.

This approach has resulted in Florida A&M University being ranked #1 for baccalaureate degrees awarded to African Americans in traditional higher education institutions and ranked #1 as the baccalaureate origin institution of African American doctorates in Natural Sciences and Engineering (NS&E), 2002-2006. The Presidential Scholars Program at Florida A&M University offers a Life-Gets-Better approach to educational attainment.

gram.) Indeed, it is in these institutions primarily that we have already invested to build capacity, so going forward it is important to build on these investments in order to improve education and research in these institutions to bolster undergraduate education in particular and graduate education in selected areas. A good example is North Carolina A&T State University, which has been reclassified as a research-intensive institution according to the Carnegie Classification System and designated an Engineering Research Center by the NSF as a result of an open national competitive process.

Hispanic-Serving Institutions

Characterizing Hispanic-Serving Institutions (HSIs) is far more difficult than describing HBCUs, or for that matter Tribal Colleges and Universities, because HSIs include a broad range of both public and private institutions, four-year colleges and universities, and community colleges. They include campuses of the University of Puerto Rico, which serve a Hispanic population in a very specific geographic location in which Hispanics are in the majority. In the continental United States, HSIs were not created to serve a specific population, as was the case for HBCUs, but rather evolved because of their geographic proximity to Hispanic populations. So, with the exception of three HSIs (Boricua College, National Hispanic University, and Hostos Community College), HSIs do not have charters or missions that address goals for educating Hispanics. They are also an "open set" to which new institutions are being added each year as they reach the federally defined 25 percent Hispanic enrollment threshold.

These institutions, therefore, vary greatly with regard to their focus on Hispanic students and how effective they are in carrying out a mission of engagement with and success for Hispanics. If our purpose in identifying the top producers of undergraduates who go on to earn PhDs is to replicate especially successful programs for Hispanics, we should consider HSIs and PWIs that are top producers as similar in that, unlike HBCUs and Tribal Colleges and Universities, HSIs are not in general using very different strategies than top-producing PWIs. Some institutions, however, which have a longer history of deep engagement with Hispanic students, may be more similar to HBCUs than more recent additions to the HSI set and, as argued in a recent report from the University of Southern California, may be more able to provide useful models for both predominantly white institutions and new or emerging HSIs to increase their productivity of STEM degree completion.[10]

We should not lose track of the fact that more than half of HSIs are two-year colleges and that over half of the Hispanic postsecondary population (including S&E majors) begin at two-year colleges. Similarly, with a few exceptions, almost all tribal colleges are two-year institutions. These facts suggest that one of the strongest recommendations we can make to increase the numbers of Hispanic and Native American scientists and engineers is to improve the transfer rates of science and engineering majors from these institutions to four-year colleges.

---

[10] Alicia C. Dowd, Lindsey E. Malcom, and Estela Mara Bensimon. 2009. *Benchmarking the Success of Latina and Latino Students in STEM to Achieve National Graduation Goals.* Los Angeles, CA: University of Southern California.

Tribal Colleges and Universities

Tribal Colleges and Universities (TCUs), on the other hand, do have a unique mission. Similar to HBCUs, TCUs were established by Tribal governments to serve American Indians, who have been historically underserved by mainstream colleges and universities, and they have been effective in preparing students for higher education and the workforce. For example, enrollment at TCUs increased by 26.8 percent for American Indian men and 17.3 percent for American Indian women between 2000 and 2005. They have been successful also in producing the largest pool of American Indian students who later complete PhDs in science and engineering. In 2005, TCUs awarded 1,662 associate's degrees, 203 bachelor's degrees, and 10 master's degrees, primarily to American Indians and predominantly to women.[11] The top producers of American Indian doctorates are in Arizona, Oklahoma, and North Dakota, all of which are in close proximity to the TCUs and reservations.

A National Academy of Engineering letter report from the steering committee for engineering studies at TCUs documents the advantages of these institutions. "TCUs offer culturally responsive education that includes cultural literacy, self-reflective analysis of attitudes and beliefs; caring, trusting, and inclusive classrooms; respect for diversity; and a transformative curriculum that engenders meaning."[12] These are especially important for children raised and educated on reservations with strong ties to tribal communities. "TCUs also offer place-based education—programs that explicitly connect students with indigenous knowledge and ways of knowing and help them discover the relationship of this knowledge to modern science and social studies."

Community Colleges

Community colleges—where the majority of underrepresented minority students begin postsecondary study—provide educational opportunity for underrepresented minority students who seek to stay in their communities, save on educational expenses, or benefit from smaller class sizes or remedial work during their first two postsecondary years. With regard to STEM, these institutions provide a variety of educational opportunities. They provide technician training or science courses for those, such as nurses, who are preparing for the workforce upon completing a two-year program. They also provide underclass coursework for engineering students who transfer to

---

[11] M. Ryu. 2008. Minorities in Higher Education. Washington, DC: American Council on Education.

[12] National Academy of Engineering. 2006. *Engineering Studies at Tribal Colleges and Universities*. Washington, DC: The National Academies Press, p. 18.

four-year programs or science courses for students who transfer to four-year institutions and either major in science or in a pre-professional program, such as nursing, dentistry, or medicine.[13] (See Box 7-4 for an example of a program promoting STEM education at Miami Dade College.) Community colleges act as a bridge to four-year institutions and should be the place to institute transition programs; they also reach out in the other direction as well, to work with K-12 through articulation agreements, summer bridge programs, and individual outreach to area high schools, assisting in the transfer from high school to college. To facilitate and increase the successful transfer of underrepresented minorities to four-year institutions, an increased emphasis on, and support for, mentoring, academic and career counseling, peer support, and undergraduate research at two-year institutions is recommended.

LEADERSHIP

Leadership in identifying and articulating minority participation and success as an institutional goal is essential at all levels for all stakeholders: the federal government, state and local governments, employers, philanthropy, professional societies, educational institutions, programs, faculty, and students. For each higher education institution that must now take action, the academic leadership—regents, trustees, presidents, provosts, deans, and department chairs—must articulate underrepresented minority participation as a key commitment both in the institutional mission and in everyday affairs in order to set a tone that raises awareness and effort. With institutional rewards connected to this mission, deeper effort and impact can be further realized. Faculty are important in the production of diversity in the student population—particularly at the PhD level—and faculty buy-in is essential.

Stakeholders must be more aggressive in investing in the development of underrepresented minority teachers, faculty, and administrators who can serve as both role models and leaders. As discussed earlier, while more minorities are receiving doctorates in science and engineering, the percentage of STEM faculty who are underrepresented minorities is very low(see, e.g., Nelson, 2007).[14] The Preparing Future Faculty Program is a national model for preparing aspiring graduate students for academic careers.[15]

---

[13] National Research Council. 2005. Enhancing the Community College Pathway to Engineering Careers. Washington, DC: The National Academies Press.

[14] See also Donna J. Nelson, A National Analysis of Minorities in Science and Engineering Faculties at Research Universities. October 31, 2007. http://chem.ou.edu/~djn/diversity/Faculty_Tables_FY07/FinalReport07.html (accessed February 25, 2009).

[15] http://www.preparing-faculty.org.

BOX 7-4
Windows of Opportunity, Miami Dade College

Windows of Opportunity is a scholarship program that assists academically promising, low-income students in obtaining the associate in arts or associate in science degrees in science, technology, engineering, or mathematics (STEM) at Miami Dade College (MDC). At least twenty-five freshman and sophomore level students participate in the program each year. Upon completion of the program, students are able to transfer to an upper division school or enter the workforce directly in their chosen field. This collaboration among eight MDC campuses brings together a diverse and experienced group of educators, business partners, and students. Program participants receive scholarships, mentoring by STEM faculty, intense academic and career planning activities, interactions with STEM professionals on and off campus, and internship experiences. The program evaluation encompasses student achievement, retention and graduation rates compared to nonprogram participants, as well as student and faculty surveys each semester, and a final student exit survey. The project is disseminated nationally by presentations of strategies, best practices, and student success rates. The program's Web portal is also publicly accessible. Upon completion of the program, participants help fill the critical shortage of scientists and engineers in Miami-Dade County. Participants not only make a contribution to South Florida, but also serve as role models to future STEM students.

In addition, a champion at the program level providing leadership dedicated to long-term improvement is typically critical to the success of programs focused on increasing the participation of underrepresented minority students. This person should be a faculty member who has the respect, power, clout, and ear of the administration. It also helps if this person is an underrepresented minority, as this provides credibility with both the majority and minority communities, and this person may then also serve as a role model to underrepresented minority students. This person is needed to organize and energize the program and obtain buy-in from other stakeholders. A person with institutional clout will bring extra resources to the program. Indeed, programs need deeper institutional buy-in for long-term sustainability; otherwise, the loss of a program champion can lead to program decline.

DEVELOPING A PROGRAM

The literature on best practices for increasing minority participation in STEM education provides guidance for the development and execution of the policies and programs that are designed to change the academic culture and sustain programs so as to encourage student retention, persistence, and completion. Below are key elements for developing a program that are necessary to transform goals into reality (BEST, 2004; Chubin and Ward, 2009; Hrabowski, 2004; Hurtado, et al. 1999; Hurtado et al., 2008; Malcom, 2004, Malcom, Chubin, and Jesse, 2004; Martin and Pearson, 2004; NRC, 2005).[16]

Resources and sustainability: The development of programs to stimulate student interest and success in STEM, both in general and for programs that target minorities, requires substantial and sustained resources. These resources provide institutional infrastructure, salaries of faculty and administrators, seed capital for the development of programs, and student financial support. For long-term sustainability of successful programs, resources need to be continual, certain, steady, and sufficient. Program success is often dependent on external support for program launch, institutional buy-in and support with time, and the development of diverse sources of funding to ensure continuity if any one piece of support is terminated.

Coordination and integration: Coordination and integration of efforts can make the aggregate of individual programs greater than the sum of their parts. This coordination and integration can be accomplished at several levels. First, as discussed earlier, federal agencies and other funding organizations can coordinate their efforts to both avoid unnecessary duplication of program support and to ensure that investments and the programs supported by those investments complement each other in a way that builds capacity and maximizes impact. Second, many programs even on the same campus operate in isolation from other efforts. Making the aggregate of individual programs greater than the sum of their parts can be accomplished by connecting program leaders to a network of such individuals who administer minority programs at the institution and in their discipline through support, information sharing, and strategic coordination.

Focus on the pipeline, career pathways, and transition points: A corollary to coordination and integration is for programs and strategies to focus on career pathways and pipeline transition points. Martin and Pearson (2004) noted that minorities suffer from high rates of attrition at each critical transition point along the pipeline from pre-K-12 all the way to the workforce. The identification and strengthening of transition points along the STEM pipeline and exposing students to career options are as important as viewing programs not as separate efforts but as pieces of larger efforts designed to move stu-

---

[16] Hurtado et al. 1999; BEST 2004; Malcom, Chubin, and Jesse 2004; Hrabowski 2004; Malcom 2004, Martin, and Pearson 2004; NRC 2005, Hurtado et al. 2008; and Chubin and Ward 2009.

dents from one step to the next, recognizing that the sequence is not linear. One promising approach may be the establishment and funding of centers of excellence that address multiple aspects of the STEM pipeline.

Program design: A successful program may be innovative or replicative and will draw on the lessons of best and worst practices in program development and implementation but be tailored to its particular institutional and disciplinary context. The components of a program will vary depending on its target population (e.g., educational stage, skills, and knowledge of students), goals, and resources. A program may include many approaches, such as outreach and recruitment, improved curricula, advanced courses, engaged mentors, peer support, research experiences, bridging, and student financial support. The program design must ensure congruence between planned goals and actual outcomes with intermittent measures to gauge short-term progress and longitudinal tracking to document impact.

Program execution: Little discussed in the literature but critical to success is program execution. Even if a program is well designed, well resourced, and appropriately targeted, without proper execution it has little chance of full success. Execution is complicated. The program requires excellent management as well, so that program components are coordinated, program administration is effective, and the program can meet or exceed its intended goals.

Program evaluation: Whether a program meets or exceeds its goals is subject to examination. Programs designed to increase the participation of underrepresented minorities benefit themselves and others by engaging in ongoing, constructive evaluation. Formative evaluations that provide feedback to programs on their design, processes, and outcomes can help those programs adjust in real time, making continuous improvements that increase impact. Summative evaluations of programs can likewise provide feedback to those programs but also make judgments about practices that can provide lessons to others.

Knowledge sharing: A corollary to the importance of program evaluation is the dissemination of information about practice that is derived from these evaluations and from other research. Successful programs draw on the lessons learned from both best and worst practice—both successful and unsuccessful programs. The Academic Competitiveness Council recommended a "living inventory" of STEM education programs that provides shared knowledge of effective practices.[17] The development and mainte-

---

[17] Irma Arispe. Presentation to the Committee. June 11, 2008.

nance of a database or clearinghouse of information from evaluation and research could enhance the availability of evidence-based approaches to formulating programs and strategies and would by extension significantly enhance effectiveness.

PROGRAM CHARACTERISTICS

While many strategies for academic and social support and integration apply equally to students in STEM fields regardless of their racial or ethnic background, for underrepresented minority students these opportunities can be critical for opening doors that would not exist for them otherwise, because they have not, on average, had the same exposure to information and experiences that facilitate movement along the STEM pathway.[18] See Box 7-5 for a list of selected promising programs. Key program characteristics that help mold the identities and motivation of students as STEM practitioners and also develop their knowledge and skills include:

- **Summer Programs:** Summer programs in mathematics, science, and engineering that include or target minority high school and undergraduate students provide experiences that stimulate interest in these fields through study, hands-on, active research or projects, and the development of a cadre of students who support each other in their interest. These programs may include college courses, workshops and seminars, career counseling, and social activities and have been found to "have positive effects on persistence (Ackermann, 1991; Garcia, 1991; Gold, Deming, and Stone, 1992; Pennick and Morning, 1983) "as well as positive effects on academic skills, test scores, first-year retention, and graduate rates (Evans, 1999)."[19]
- **Research Experiences:** At the undergraduate level, engagement in rich research experiences allows for the further development of interest and competence in and identification with STEM. Research has shown that these experiences are critical in convincing students to pursue graduate study in STEM disciplines. They provide experience with the operations of science, very often seize the interest of students who then develop a fascination that translates into a career in STEM (Bauer and Bennett, 2003; Chubin and Ward, 2009; Clewell et al., 2005; Hackett, Croissant, and Schneider, 1992; Highsmith, Denes, and Pierre, 1998; Hunter et al., 2007; Kardash, 2000; Lopatto, 2003, 2004, 2007; Nagda et al., 1998; NRC, 2005a ; NRC, 2009; Pascarella and Staver, 1985; Rueckert, 2002; Russell et al., 2007; Walters, 1997).

---

[18]B. C. Clewell et al. 2006. *Final Report on the Evaluation of the National Science Foundation Louis Stokes Alliances for Minority Participation Program.* Washington, DC: The Urban Institute, pp. 34-35.

[19] Ibid, p. 22.

- **Professional Development Activities:** Similar to the importance of enriching research experiences, the provision of opportunities for undergraduate and graduate students to engage in professional development activities, particularly in graduate programs, will provide additional opportunities to develop and socialize students within a discipline and profession. These activities include opportunities for networking, participation in conferences, and presentations of research (on campus or in other professional settings).[20]

- **Academic Support and Social Integration:** Even if students are prepared, have adequate information, and are ambitious and talented enough to succeed in STEM fields, success may also hinge on the extent to which students feel socially and intellectually integrated into their academic programs and campus environments. The importance of social and intellectual integration for success is critical to all students, regardless of background. For minority students, who may feel like outsiders because they see few others "like themselves" among the student and faculty populations, this issue takes on even greater salience. The development of peer-to-peer support, study groups, program activities fostering social integration, and tutoring and mentoring programs may go a long way to overcome this critical hurdle. (See Box 7-6 for a review of the literature on academic and social support activities.)

- **Mentoring:** Engaged mentors can provide students with information, advice, and guidance and support both generally and at critical decision points. This kind of support helps undergraduate and graduate students take full advantage of a program and may be the difference between a student completing or leaving a program. At the undergraduate level, helping a student prepare and apply for graduate school can make the difference between whether a minority student continues in the STEM pathway. In graduate school, mentors provide important guidance and support to students, reducing attrition, helping students maximize their educational experience, and providing guidance on launching a career.[21] It is often suggested that students are best served by mentors from the same minority groups as the students, especially minority professionals, but we note the "countless others" who have served as excellent mentors and must also do so in the future.

Beyond these student-focused activities, there are additional steps that institutions and STEM departments can engage in that make a difference in student outcomes. The availability or accessibility of institutional research infrastructure—that is, laboratories and equipment—and provision for

---

[20] National Research Council, 2005. *Assessment of NIH Minority Research and Training Programs.* Washington, DC: The National Academies Press.

[21] National Research Council. 2005. *Assessment of NIH Minority Research and Training Programs.* Washington, DC: The National Academies Press.

BOX 7-5
Selected Promising Interventions

These programs are listed because of their prominence and, in some instances, longevity. Some have undergone rigorous evaluations and shown to be effective, while others have more anecdotal reports of their impact.

Undergraduate Programs

Meyerhoff Scholars Program
The University of Maryland Baltimore County Meyerhoff Scholars Program is a leading producer of high-achieving minorities who go on to graduate and professional study and careers in STEM. Program components include summer bridge program, scholarship support, tutoring and mentoring, research experiences, study abroad, and networking.

Mathematics Workshop Program
The program was developed by Uri Treisman of the University of California, Berkeley, to reverse the low success rate in entry-level calculus and the high attrition rate in math-related fields for minority students who were interested in pursuing STEM careers.

Research Initiative for Scientific Enhancement Program (RISE)
RISE enables minority institutions to increase the number of minorities in biomedical and behavioral research who complete PhDs in these fields. It is funded by the National Institutes of Health.

The Leadership Alliance
Based at Brown University, the Leadership Alliance is an academic consortium of 33 diverse institutions dedicated to promoting the entry of minorities into graduate school and the professoriate. Features include summer research, faculty resource network, specialized seminars and handbooks, graduate student support, and a national symposium.

Louis Stokes Alliances for Minority Participation (LSAMP)
Established by the National Science Foundation, the LSAMP program aims to develop strategies to increase the quality and quantity of minority students who successfully complete degrees in STEM through multi-institution alliances across the nation.

Howard Hughes Medical Institute Exceptional Research Opportunities Program (EXROP)
EXROP provides talented undergraduates from disadvantaged backgrounds with summer research experiences in the labs of HHMI investigators and HHMI professors. The students are selected by HHMI professors ad invited directors of HHMI-funded undergraduate programs at institutions. They attend meetings at HHMI headquarters, where they present their research in a poster session, network with their peers and HHMI scientists, and hear from scientific researchers from various backgrounds and at various stages in their career.

Graduate Programs

The National Consortium for Graduate Degrees for Minorities in Science and Engineering (GEM)

The GEM Consortium leverages its base funding and resources of consortium member universities to combine paid fellowships and internships with industry to prepare minorities for the global workplace. Through its university and employer members and other partners, GEM awards portable graduate fellowships and builds mentor networks for recipients.

National Institutes of Health (NIH) Minority Research and Training Programs

The (R25) Bridge to the Doctorate, (T32) NRSA Minority Institutional Research Training program, (T35) NRSA Short-Term Institutional Training Grants, (F31) NRSA Minority Institutional Research Training and Predoctoral Fellowship Programs, and (R03) Minority Dissertation Research Grant all help to increase the supply of minorities in graduate programs.

Ronald E. McNair Post-Baccalaureate Achievement Program

McNair is one of eight TRIO programs funded by the Department of Education. Funds are awarded to institutions to prepare underrepresented students to complete the PhD. Project activities include support for scholarly activities, summer internships, mentoring, and financial support.

Alliances for Graduate Education and the Professoriate (AGEP)

This program furthers the graduate education of underrepresented STEM students through the doctorate level, preparing them for fulfilling opportunities and productive careers as STEM faculty and research professionals. AGEP also supports the transformation of institutional culture to attract and retain STEM doctoral students into the professorate.

Louis Stokes Alliances for Minority Participation (LSAMP) Bridge to the Doctorate

The NSF LSAMP Bridge to the Doctorate provides two years of fellowship support for graduate students in STEM disciplines. Awards include student stipends and a cost-of-education allowance to the institution for tuition, health insurance, and other normal fees.

Ford Foundation Fellowships Program

Administered by the National Research Council, the program provides graduate fellowships to minorities who are committed to an academic career in teaching and research.

SOURCE: Cheryl B. Leggon and Willie Pearson, Jr. 2008. Assessing programs to improve minority participation in STEM Fields: What we know and what we need to know, in R. Ehrenberg and C. Kuh, eds., *Doctoral Education and the Faculty of the Future.* Ithaca, NY: Cornell University Press.

BOX 7-6
Review of Literature on Student Support

Tutoring. Tutoring has been shown to be effective in increasing student persistence, positive attitudes toward subjects, and student performance (Carman, 1975; Gahan-Rech, Stephens, and Buchalter, 1989). No differences in achievement outcomes have been found for peer tutoring versus staff tutoring (Moust and Schmidt, 1994). Benefits of tutoring have been established not only for those receiving tutoring, but also for the tutors themselves (Bargh and Schul, 1980; Good, Halpin, and Halpin, 1998).

Peer Study Groups. Peer study groups are a fundamental component of several successful programs to increase the achievement and retention of underrepresented minorities in STEM. Program evaluation results of both the Mathematics Workshop Program (MWP) and replication programs have shown that workshop participants who work in peer study groups are more likely to persist in SME, graduate, and earn high grades in the study subject (Alexander, Burda, and Millar, 1997; Bonsangue and Drew, 1995; Fullilove and Treisman, 1990; Moreno and Muller, 1999; Murphy, Stafford, and McCreary, 1998; Treisman 1992). A meta-analysis of the effects of small-group learning on undergraduate STEM students found that various forms of small-group learning are effective in increasing academic achievement, persistence in STEM, and developing more favorable attitudes toward learning (Springer, Stanne, and Donovan, 1999).

Skills-Building Seminars. The effectiveness of seminars and workshops to build study skills, test-taking strategies, time management, and other skills that are useful to college success has been rarely studied (Gándara, 1999), although limited evidence that they are effective has been found (Novels and Ender, 1988).

Learning Centers. There is not much research on the effects of learning centers, but observations linking their presence on campus to student learning have been documented (Holton and Horton, 1996).

Academic Advising and Mentoring There have been several studies of academic advising as a strategy used in retention programs. Of these, some studies have established their positive effect on student retention or satisfaction with their institutions (Backhus, 1989; Forrest, cited in Pascarella and Terenzini, 1991; Lowe and Toney, 2001; Trippi and Cheatham, 1991). Although mentoring has become an important element of most intervention programs for underrepresented minorities, research evidence on its effectiveness continues to be mostly qualitative. What evidence does exist, however, suggests that for minority students, mentoring results in such positive outcomes as higher GPAs, lower attrition, increased self-efficacy, and better defined academic goals (Santos and Reigadas, 2002; Schwitzer and Thomas, 1998; Thile and Matt, 1995). Mentoring has been said to facilitate students' academic and social integration (Redmond, 1990).

SOURCE: B.C. Clewel, et al. 2006. *Final Report of the Evaluation of the Louis Stokes Alliances for Minority Progress Program*, pp. 38-39.

research opportunities at federal laboratories may determine the level of success in STEM areas that students may aspire to. Institutions should procure adequate facilities and equipment or partner as possible with other nearby institutions to facilitate the access of their students to these other resources. The federal government can assist by providing institutions with funding for facilities and equipment or by supporting the development of networks among institutions that would provide access to these types of resources, among other things.

In addition to equipment and facilities, the curriculum may also require a makeover. As shown in the Higher Education Research Institute data displayed in Chapter 4, undergraduates regardless of race/ethnicity are less likely to persist and complete in their intended major if they begin in a STEM field compared to a non-STEM field. Seymour and Hewitt (1997) found that students switched out of mathematics, science, and engineering majors at higher rates than for other fields and that this was due in part to the culture of these fields and the characteristics of classes, particularly introductory classes, in these fields, some of which sought intentionally to "weed out" students. Further, they discovered that women and underrepresented minorities were more likely to be turned off by the way science is taught, internalizing difficulties when facing challenges rather than assigning blame to the larger scientific and educational culture. Seymour and Hewitt found that students' experiences were characterized by:

- Poor teaching or organization of material;
- Hard or confusing material, combined with loss of confidence in their ability to do science;
- Cutthroat competition in assessment systems geared more to weeding out than to encouraging students;
- Dull subject matter; and
- Grading systems that did not reflect what students felt they had accomplished.

Further, many of those who stayed also complained about poor teaching and an unpleasant atmosphere. Both male and female switchers reported that negative experiences in freshman science were more important than positive experiences in other fields in reaching their decision to leave. Efforts on the part of institutions, departments, and faculty to change curricula to provide more hands-on, active learning and to encourage rather than weed out students could play a significant role in increasing the participation of underrepresented minorities.[22]

---

[22] Elaine Seymour and Nancy M. Hewitt. 1997. *Talking About Leaving: Why Undergraduates Leave the Sciences.* Boulder, CO: Westview Press, pp. 10-11.

8

Recommendations and Implementation Actions

As described in previous chapters, a successful national effort to increase the participation and success of underrepresented minorities in STEM must be urgent, sustained, comprehensive, intensive, coordinated, and informed. It must also cut across all educational stages and stakeholder groups. With these principles in mind, the committee has developed six broad recommendations and a description of actions that should be taken by specific stakeholders. Following the six broad recommendations, the committee proposes two top priorities that should serve as the near-term focal point for national policies for broadening participation.

PREPARATION

Recommendation 1: *Preschool Through Grade 3*

Prepare America's children for school through preschool and early education programs that develop reading readiness, provide early mathematics skills, and introduce concepts of creativity and discovery.

Federal Government

• The federal government should fully fund Head Start and pre-kindergarten school-readiness programs. The American Recovery and Reinvestment Act provided a one-time infusion of $1.1 billion to double the number of children served by Early Head Start over two years and an additional $1 billion to expand and improve Head Start. This level of funding should be sustained and expanded, working toward the goal of voluntary, high-quality preschool education, universally available to 3- and 4-year old children from qualifying families.

• The federal government, in coordination with states and local school districts, should consider targeted resources to perpetuate gains obtained through Head Start and public pre-K programs once students are enrolled in elementary school; incentives for states to expand the capacity and improve the quality of public pre-K programs; and the increased integration of child care, Head Start, and state pre-K programs to reduce the disparities in early education and school readiness.

• The federal Head Start program, in conjunction with the Department of Education and National Science Foundation, should provide dissemination and training on curricular tools for Head Start and public pre-K programs that facilitate the introduction of scientific skills, such as observing nature, formulating questions, and creativity.

State Governments

• We echo the recent recommendation of the College Board (2008a) that "Governors and legislators, working with educators, community groups, and experts on Head Start and early childhood education, should develop funding formulas to help communities establish and create effective preschool programs and standards for their operation."[1]

• State systems should work with educators and experts to align early childhood programs with public school curriculum and quality standards, including those for mathematics and science, to ensure the successful matriculation of children during the early grades.

Local School Districts

• Local school districts should offer guidance on how to align preschool curricula with learning expectations in kindergarten.[2] Experts suggest aligning preschool curricula with expectations through 3rd grade.

• Local school districts should target resources to perpetuate gains obtained through preschool programs once students are enrolled in elementary school by adopting promising practices and proven interventions.

---

[1] College Board. 2008. *Coming to Our Senses: Education and the American Future.* New York, NY: The College Board.

[2] College Board. 2008. *Coming to Our Senses: Education and the American Future.* New York, NY: The College Board.

Recommendation 2: *K-12 Mathematics and Science*

Increase America's talent pool by vastly improving K-12 science and mathematics education for underrepresented minorities.

Federal Government

- The federal government should increase Title I funding, require equitable state and district budgeting practices for schools whose populations include high proportions of students from economically disadvantaged families, and require districts to publicly report per-student expenditures by funding source (state, local, and federal) and by school. As minority students are overrepresented among such families, this will help to narrow the academic achievement gaps overall.[3]

- The federal government should reform the No Child Left Behind (NCLB) Act by improving its overall effectiveness, especially for schools with large minority populations. As part of this reform, the Act should support states, school districts, and schools that identify and mitigate gaps (between and within schools) in student performance in English language skills, mathematics, and science. The reform of NCLB should retain the requirement that schools be held responsible for the achievement of the various subgroups of students they serve by continuing to report test scores disaggregated by race/ethnicity.

- The federal government should expand its programs that impact K-12 science and mathematics education (e.g., Mathematics and Science Partnerships programs at the National Science Foundation and U.S. Department of Education) in order to enhance schools' capacity to provide challenging curricula for all students; contribute to a greater understanding of how students effectively learn mathematics and science and how teacher preparation and professional development can be improved; engage and support scientists, mathematicians, and engineers at local universities and local industries in working with K-12 educators and students; and promote institutional and organizational change in education systems—from kindergarten through graduate school—to sustain partnerships' promising practices and policies.

- The federal government should seek to improve early intervention programs such as the TRIO programs, especially the Upward Bound Math-Science program, and augment budgets as warranted.

---

[3] D. Hall and N. Ushomirsky. 2010. *Close the Hidden Funding Gaps in Our Schools.* Washington, DC: The Education Trust.

State Governments

• States should adopt evidence-based curriculum standards across subject areas, including science and mathematics, to ensure college and career readiness for all students. For example, the National Governors Association Center for Best Practices and the Council of Chief State School Officers have drafted Common Core Standards in English-language arts and mathematics for grades K-12 to provide a clear and consistent framework to prepare our children for college and the workforce.[4] They are aligned with college and work expectations, include rigorous content and application of knowledge through high-order skills, informed by international benchmarks, and evidence-based.

• As a corollary, states should develop rigorous testing programs that identify performance gaps—including those for science and mathematics—and provide resources, strategies, and programs to address them.

• States should provide equitable resources and quality teachers to schools with high minority populations.

• States should support the establishment of magnet high schools for science and mathematics in each major jurisdiction within a state and insist that these schools make inclusivity a requirement.

Local School Districts

• School districts should develop programs that identify and encourage minority students to more fully develop their knowledge base and potential in mathematics and science. These programs should include efforts to encourage minority students to enroll in and pass Advanced Placement (AP), International Baccalaureate (IB), or similar advanced courses and examinations.

• Local school districts should develop and provide quality mathematics and science curricula that include active, hands-on, project-based learning that improve understanding of science and scientific processes. These may be augmented through informal education programs such as those at science centers and museums.

• Schools and teachers should capitalize on the findings of research on students who are low achievers, have difficulties in mathematics, or have learning disabilities related to mathematics in order to improve instruction for them. This research tells us that the effective practice includes:[5]

— Explicit methods of instruction on a regular basis,
— Clear problem-solving models,

---

[4] Common Core Standards Initiative, http://www.corestandards.org.

[5] National Mathematics Advisory Panel: Foundations for Success: Report of the National Mathematics Advisory Panel (from presentation to committee by Irma Arispe, White House Office of Science and Technology Policy, June 11, 2008).

— Carefully orchestrated examples/sequences of examples,
— Concrete objects to understand abstract representations and notation, and
— Participatory thinking aloud by students and teachers.
* Local school districts should encourage teachers and administrators to hold genuinely high expectations for minority students and follow through on activities and programs that help students meet those expectations.

Nonprofits

* Nonprofits should continue to pioneer new program approaches that employ innovative strategies or target particular niches. Nonprofit groups, with support of federal agencies or philanthropic organizations, have developed promising programs that may yield significant impacts. These kinds of programs should be fostered so that we can continue to improve our means for reaching underrepresented minority students and engaging them in mathematics and science. The Algebra Project provides an example of a program established to target minority students, encourage them to demand access to quality mathematics instruction and use this as a springboard for college and beyond. Similarly, the Center for the Advancement of Hispanics in Science and Engineering Education offers rigorous educational and leadership development programs for 5th graders and beyond to improve students' future performance in STEM programs.

Recommendation 3: *K-12 Teacher Preparation and Retention*

Improve K-12 mathematics and science education for underrepresented minorities overall by improving the preparedness of those who teach them those subjects.

Federal Government

* The federal government should provide incentives for the annual recruitment, retention, and professional development of science and mathematics teachers who teach minority students. *Rising Above the Gathering Storm* recommended that the federal government "annually recruit 10,000 science and mathematics teachers by awarding 4-year scholarships." As minority students comprise 36.6 percent of K-12 students in the United States (as of 2006), 3,660 of these new science and mathematics teachers should, upon graduation, be allocated to schools with a predominantly minority enrollment. Minority math and science teachers should be declared an area of national need.

- The federal government should strengthen the capacity of teacher education programs at minority-serving institutions to prepare and produce quality math and science teachers who intend to teach underrepresented minority students.
- Drawing on program evaluations and policy development, the federal government should improve the quality of its suite of teacher preparation programs across federal agencies. These programs, whether targeted toward minorities or not, should include a special focus on increasing the number, quality, and diversity of mathematics and science teachers, especially in underserved areas.

State Governments

- States should, along with state colleges and universities, coordinate STEM teacher training programs that recruit, prepare, and place qualified teachers in high-needs schools proportionately to all other schools. An emphasis should be placed on reducing the use of out-of-field science and mathematics teachers in high-minority schools.
- States should provide incentives for qualified teachers to work in schools and districts with high-minority and low-income enrollments and seek to reduce turnover among these teachers.

Higher Education Institutions

- Higher education institutions should increase the recruitment, preparation, professional development, and retention of well-qualified elementary and secondary teachers in mathematics and the sciences who are prepared to teach diverse students. This preparation should include the requirement that the core teacher education curriculum provide courses in multicultural approaches to pedagogy.

POSTSECONDARY SUCCESS

Recommendation 4: *Access and Motivation*

Improve access to all postsecondary education and technical training and increase underrepresented minority student awareness of and motivation for STEM education and careers through improved information, counseling, and outreach.

Federal Government

• The federal government should ensure that outreach programs linking postsecondary institutions and K-12 schools and systems, such as the Graduate Teaching Fellows in K-12 (GK-12), Opportunities for Enhancing Diversity in the Geosciences, and the Mathematics and Science Partnerships (MSP) programs at the National Science Foundation and the U.S. Department of Education, include a strong emphasis within them of improving K-12 mathematics and science education and awareness for underrepresented minority students. This may entail greater coordination across outreach programs in the same institution or geographic region.

• The U.S. Department of Education must improve the efficiency and effectiveness of the TRIO Upward Bound program, which has the potential for a strong positive impact on underrepresented minority enrollment in college and for furthering minorities' aspirations to major in STEM.

States and School Districts

• As a standard, states must require middle and high schools to have at least one counselor for every 250 students and charge these counselors with providing students with a "robust college counseling program."[6] Within this counseling program, counselors should follow the lead of mathematics and science teachers to encourage interested and motivated students to pursue STEM education and careers and provide them with information about the course prerequisites for success in STEM education in college.

• States and school districts should introduce students to STEM careers, starting in preschool, through awareness activities and informal education programs that would include speakers (role models), activities, field trips, participation in science or engineering programs, and links to summer programs.[7] These may be accomplished in partnership with employers and nonprofit organizations. These must include an emphasis on and programs targeted to increasing the participation of underrepresented minorities.

Employers

• Businesses, government agencies, and higher education institutions should work to plant the seeds of interest in STEM by allowing staff to visit elementary schools where they can discuss science and engineering and

---

[6] College Board, *Coming to Our Senses.*

[7] National Action Council for Minorities in Engineering. 2008. *Confronting the "New" American Dilemma, Executive Summary*, p. 8.

talk with students about and encourage them to consider STEM careers.[8]
To the extent possible, when such staff are of the same racial/ethnic group
as the students they visit, they can provide role models, particularly for
underrepresented minority students, that they may not encounter often in
their communities.

• Employers also should offer on-site internships to underrepresented
minority students and teachers and provide access to resources such as the
latest equipment and software.[9]

Higher Education Institutions

• Higher education institutions should engage in targeted outreach
and recruitment activities that could constitute a "feeder system" to help
cultivate underrepresented minority students who may aspire to enroll in
these institutions.[10]

• Higher education institutions should develop summer programs
in mathematics, science, and engineering that include or target underrepresented minority high school students. These programs should provide
experiences that stimulate interest in these fields through study and hands-
on research or projects and that develop a cadre of students who support
each other in their interests.

Recommendation 5: *Affordability*

*Develop America's advanced STEM workforce by providing adequate
financial support to underrepresented minority students in undergraduate
and graduate STEM education.*

Federal Government

• In addition to supporting underrepresented minorities through
need-based college financial aid programs (e.g., Pell Grants), the federal
government should provide financial support to underrepresented minorities for participation in undergraduate STEM programs across institution
types: community colleges, minority-serving institutions, and majority-
serving institutions. *Rising Above the Gathering Storm* recommended that

---

[8] National Action Council for Minorities in Engineering. 2008. *Confronting the "New"
American Dilemma, Executive Summary*, p. 8.

[9] NACME. 2008. *Confronting the "New" American Dilemma, Executive Summary*, p. 8.

[10] Daryl E. Chubin and Wanda E. Ward. 2009. *Building on the BEST Principles and Evidence: A Framework for Broadening Participation*, in Mary K. Boyd and Jodi L. Wesemann,
eds., Broadening Participation in Undergraduate Research: Fostering Excellence and Enhancing the Impact, Washington, DC: Council of Undergraduate Research, pp. 21-30.

the federal government "increase the number and proportion of U.S. citizens who earn bachelor's degrees in the physical sciences, the life sciences, engineering, and mathematics by providing 25,000 new 4-year competitive undergraduate scholarships each year to U.S. citizens attending U.S. institutions." Federal efforts to implement this recommendation should include an emphasis also on increasing participation of underrepresented minority students. We recommend that the federal government make a strong effort to encourage underrepresented minorities to apply for and participate in this program as, at a minimum, at least 40 percent of these scholarships will need to be attained by underrepresented minority students for the nation to make any progress toward increased underrepresented minority retention and completion in undergraduate STEM programs.

• *Rising Above the Gathering Storm* recommended that the federal government "Increase the number of U.S. citizens pursuing graduate study in 'areas of national need' by funding 5,000 new graduate fellowships each year." Again, federal efforts to implement this recommendation should include an emphasis on also increasing participation of underrepresented minority students in all types of institutions of higher education, particularly research universities, where underrepresented minorities must have equitable representation in the student body and faculty if they are to fully contribute to our nation's research and take part in national STEM leadership. The ideal package—particularly at the graduate school level—would allow the student to focus on studies and research full-time, without increasing debt burden or working in a non-STEM related job off-campus that would be a distraction.

• The federal government—along with other stakeholders—should increase funding for undergraduate and graduate STEM programs focused on increasing the participation and success of underrepresented minority students through engaged mentoring, enriching research experiences, and opportunities to publish, present, and network. To the extent that students can participate in conferences, present papers, engage in summer research, or take advantage of similar activities, the deeper their commitment to their program, their discipline, and their profession. Students from disadvantaged backgrounds will likely require additional financial support for these activities as well. Sources of this support may include institutional funds or funding from federal or philanthropic programs.

• The federal government should increase funding for the operating expenses of TCUs and increase the level authorized under the Tribally Controlled College or University Assistance Act of 1978.

• The federal government should assess the impact of the American Competitiveness Grant (ACG) and National SMART Grant programs to ensure they are best meeting the needs of students with potential for success in STEM. Early reports indicated that far fewer students than originally anticipated were taking advantage of the program. While use has increased, the government should review its program outreach and selection processes as well as eligibility criteria.

State Governments

- State governments must assist with the education of under-represented minority students in STEM. They may do so by more fully supporting public higher education generally in their jurisdictions. State appropriations as a percentage of the operating budgets of public institutions has been declining since the early 1970s, about the same time that civil rights efforts first helped increase diversity on our nation's public campuses in a substantial way.[11] State appropriation levels have always shifted with general economic circumstances. However, since the mid-1990s, the increases in appropriations have failed to compensate, in real terms, for the periodic downturns.

Philanthropy

- With relative freedom to explore new program approaches, foundations should develop and/or fund programs that employ innovative strategies or target particular niches in undergraduate and graduate STEM education for underrepresented minorities. For example, the Gates Millennium Scholars program, funded by a $1 billion grant to UNCF from the Bill and Melinda Gates Foundation in 1999, seeks to promote academic excellence and provide an opportunity for outstanding minority students, with significant financial need in education, engineering, library science mathematics, public health, and the sciences to reach their highest potential by reducing financial barriers and providing seamless support from undergraduate through doctoral programs. Similarly, the Howard Hughes Medical Institute (HHMI) has developed the Exceptional Research Opportunities Program to increase the number of minority doctorates by selecting HHMI grantee students to conduct research in an HHMI lab, receive continued mentoring and networking, and attend summer meetings. They are eligible for pre-doctoral fellowships, and to date one-third of participants have attained the PhD. The Ford Foundation Fellowship Program provides a third example of philanthropic support to increase the diversity of the nation's college and university faculties, to maximize the educational benefits of diversity, and to increase the number of professors who can and will use diversity as a resource for enriching the education of all students. To facilitate these goals

---

[11] Christopher Newfield. 2008. *Unmaking the Public University: The Forty Year Assault on the Middle Class*. Boston, MA: Harvard University Press.

directly, the program awards fellowships at the predoctoral, dissertation, and postdoctoral levels. Finally, the Alfred P. Sloan Foundation provides scholarships for minority students who are beginning their doctoral work in mathematics, natural sciences, and engineering and connects students to faculty and departments with demonstrated success in sending their students to doctoral programs. A special program enables American Indian master's and doctoral mathematics, natural sciences, and engineering students to apply for scholarships to attend one of five universities. The Foundation also helps position minority PhD's for faculty positions at research universities.

Higher Education Institutions

- Higher education institutions have a responsibility, particularly when they charge increasingly high tuition and fees, to provide need-based financial assistance to students and families who have a demonstrated need for such assistance. This is critical for underrepresented minorities in STEM who are from low- and moderate-income families. Harvard University and Brown University have been proactive in this regard. Harvard's financial aid initiative for low- and middle-income families reduces the contributions of families with incomes between $60,000 and $80,000; those with incomes of less than $60,000 are not expected to contribute to the cost of their children's attending Harvard. Consequently, the class of 2010 was the most diverse in Harvard's history. Brown University's financial aid initiative has similar family income thresholds, and students are able to use outside scholarships to eliminate all of the student-effort components in their financial aid awards.
- Doctoral institutions must do a much better job of including more underrepresented minorities in STEM as research assistants. Fellowships are the most prevalent form of support for underrepresented minorities. Such assistantships help bridge the gap between what they can afford to pay and the cost of attendance when other sources of support are not available or do not cover the full cost of attendance. The need for loans or outside work is negatively correlated with enrollment and completion. Furthermore, work as an RA provides skills and experience that improve educational outcomes and make graduates more competitive in the job market.

Recommendation 6: *Academic and Social Support*

Take coordinated action to transform the nation's higher education institutions to increase inclusion of and college completion and success in STEM education for underrepresented minorities.

Federal Government

- The federal government should increase funding for infrastructure, research, curriculum development, and professional training at minority serving institutions through such programs as HBCU-UP, TCU-UP, RISE, MARC/MBRS, and CREST. These grants should be larger in size so they can have a substantial impact on the recipient campuses; they should be competitively awarded to institutions that can most effectively use them to strengthen the quality of their STEM education and, therefore, the preparation of underrepresented minority students in STEM for both the workforce and competitive graduate programs; moreover, they should require rigorous program evaluation.

- Federal agencies, particularly those with large science and engineering research portfolios, should hold institutions that receive federal research funding accountable for broadening participation in STEM—particularly at the doctoral level. The National Science Foundation (NSF) asks potential grantees to explain both the intellectual merit of the proposed research as well as its broader impact. Under this evaluation criterion, the NSF should continue to emphasize the importance of broadening participation of underrepresented minorities as one critical way that research projects can achieve important broader impact goals. The National Institutes of Health (NIH) does not ask grantees for a discussion of how a proposed project will address broader impact. We recommend that the NIH reconsider the decision not to include such a criterion. All agencies should consider broader impact criteria, particularly the Department of Defense, the Department of Energy, the Department of Agriculture, National Aeronautics and Space Administration, National Institute of Standards and Technology, and National Oceanographic and Atmospheric Administration.

- The federal government should create and fund a program—based on the goals and structures of the NSF ADVANCE program—to increase the representation and advancement of underrepresented minorities in academic science and engineering careers, thereby contributing to the development of a more diverse science and engineering workforce. As described on the NSF Web site,

> ADVANCE encourages institutions of higher education and the broader science, technology, engineering and mathematics (STEM) community, including professional societies and other STEM-related not-for-profit organizations, to address various aspects of STEM academic culture and institutional structure that may differentially affect women faculty and academic administrators.

A similar program focused on underrepresented minorities could also play "an integral part of the NSF's multifaceted strategy to broaden participation in the STEM workforce" and support the critical role of the Foundation in advancing the status of underrepresented minorities in academic science and engineering.

• Federal agencies should continue to build cadres of mentors at higher education institutions nationwide. Those who have received the Presidential Awards for Excellence in Science, Mathematics and Engineering Mentoring can serve as a national resource, and federal agencies should draw on them to build knowledge regarding the identification and nurturing of talent and to help cultivate other mentors. In addition, the NSF should require a mentoring plan in program solicitations for graduate students as it does for postdocs.

Higher Education Institutions

• At the most general level, the institutional commitment to inclusiveness and the policies used to express that commitment play a critical contextual role for programs designed to increase underrepresented minority participation in undergraduate and graduate STEM. Therefore, a campus-wide commitment to inclusiveness provides the best environment for planting the seeds of diversity. This should be articulated by university leaders both in the university mission and in everyday affairs. Leadership is essential at all levels of academia—the faculty, department chairs, deans, provosts, chancellors and presidents, and even regents and trustees—for programs to work, that is, to increase the participation of underrepresented minorities in a significant way. The visible and continuing commitment of these leaders to diversity and to minority participation provides the overall, critical tone that signals appropriate actions for others.

• Institutions should further reinforce the commitment to diversity by rewarding faculty in the promotion and tenure process for developing student talent and coaching junior faculty, both in general and for underrepresented minorities. They can support this by also providing professional development opportunities for faculty in areas such as mentoring diverse students.

• Higher education institutions should continue to remove systemic barriers to the participation of underrepresented minorities in college by developing, implementing, and enforcing undergraduate and graduate admissions and financial aid policies that reinforce diversity within the legal parameters of the Michigan decisions in order to ensure a significant and sufficient overall level of minority participation on campus.

• Institutions should develop bridging programs to enable students to matriculate along the STEM education continuum. These programs include academic preparation, guidance from mentors on mastering the transition, the development of connections between programs, and financial support as necessary. Key transition points at which bridging can help include the transition from community colleges to four-year institutions, from undergraduate to graduate programs, and from master's to doctoral programs.

• Undergraduate and graduate STEM education programs should incorporate social inclusion strategies that include peer-to-peer support, study groups, program activities fostering social integration, and tutoring and mentoring programs. These strategies should be implemented as complements to summer programs, enriching research experiences, bridging programs, and professional development activities.[12]

• Higher education institutions should encourage genuinely high expectations on the part of faculty toward minority students and follow through on activities and programs that help students meet those expectations.

• Higher education institutions, especially research universities, should replicate the practices of institutions with demonstrated success in producing large numbers of minorities with STEM undergraduate and graduate degrees. A successful program will draw on the lessons of best and worst practices in program development and implementation but will be tailored to its particular institutional and disciplinary context. The long-term success of programs is often dependent on diverse sources of funding, including institutional resources, to ensure continuity if any one piece of support is terminated.

• Institutions should procure adequate facilities and equipment or partner as possible with industry, federal laboratories, and other institutions to facilitate student access to these other resources. The federal government can assist by providing institutions with funding for facilities and equipment or by supporting the development of networks among institutions that would provide access to them, among other things.

Professional Associations and Scientific Societies

• Professional associations and scientific societies should make recruitment and retention of underrepresented minority scientists and engineers an organizational goal and implement programs designed to reach that goal. These organizations should "work with their membership, academic institutions, and funding agencies to monitor the impact of programs aimed at broadening participation in science and to develop and sustain effective, new

---

[12] The Howard Hughes Medical Institute developed a symposia program in which invited participating institutions were asked to provide data on their minority programs. The data collected confirmed that underrepresented minorities were more likely to drop out of programs early, but that early intervention strategies made a difference: Such strategies include summer bridge, peer mentoring, peer leadership, coaching for social aspects, study groups, early research opportunities, and faculty mentoring.

initiatives." These organizations should "work together to communicate the importance of broadening participation in science to their members, the public and policy makers."[13]

•   Professional associations and scientific societies should implement policies and programs designed to encourage mentoring. "Mentoring underrepresented minorities should be integral to any initiative or program designed to enhance diversity in the sciences. Organizations should emphasize the importance of mentoring and promote and facilitate mentoring of students and junior scientists by their senior colleagues." Among more specific ideas, they should "reward faculty for time spent on mentoring, and encourage the provision of grants that offer protected time for mentoring activities."[14]

Industry and Federal Laboratories

•   Industry and federal laboratories can broaden the participation of underrepresented minorities in science and engineering through structured incentives and programs to ensure sustained impact, such as internships, research assistantships, scholarships, and fellowships for undergraduate and graduate students. Industry and federal laboratories also should provide greater opportunities for minority faculty research collaboration and exchanges to increase their chances for tenure and promotion considerations.

•   Industry and federal laboratories should expand their partnerships with institutions that enroll large numbers of underrepresented minorities in STEM in order to increase the articulations between universities and industry/federal laboratories and expand the population of role models to interact with an increasingly diverse student population that will become the future workforce.

•   Industry and federal laboratories can be pivotal in enhancing the research capacity of minority-serving institutions, stimulating innovation in undergraduate and graduate education, and facilitating interdisciplinary training by providing much needed laboratory equipment.

TOP PRIORITY ACTIONS

Among the recommendations and implementation actions presented, we have identified two areas of highest priority for near-term action. We chose them because we believe they can have the most immediate impact

---

[13] Consortium for Social Science Associations, *Enhancing Diversity in Science: A Leadership Retreat on the Role of Professional associations and Scientific Societies: A Summary Report*, February 28, 2008.

[14] Ibid.

on the critical transition points in the STEM education pathway for underrepresented minorities.

Priority 1: Undergraduate Retention and Completion

We propose, as a short-term focus for increasing the participation and success of underrepresented minorities in STEM, policies and programs that seek to increase undergraduate retention and completion through strong academic, social, and financial support. Financial support for underrepresented minorities that allows them to focus on and succeed in STEM will increase completion and better prepare them for the path ahead. This financial assistance should be provided through higher education institutions along with programs that simultaneously integrate academic, social, and professional development.

We have chosen this focus for several reasons:

• A cadre of qualified underrepresented minorities already exists who attend college, declare an interest in majoring in the natural sciences or engineering, and either do not complete a degree or switch out of STEM before graduating. An intense effort to reduce this attrition and bolster baccalaureate completion represents the most straightforward way to retain these students.

• An increase in the completion of undergraduate STEM degrees by this population may also have impacts up and down the pathway. The visibility of increased undergraduate success may stimulate interest in STEM on the part of younger cohorts. And the increase in overall numbers will increase the pool of underrepresented minorities who may consider graduate education and careers in STEM.

The goal is to increase participation at all types of higher education institutions, including research universities, where underrepresented minorities can contribute to research, become more prominent leaders, and serve as role models. This will fuel the pipeline of minority scientists and engineers in the STEM workforce.

Between 1998 and 2007, the number of underrepresented minorities earning bachelor's degrees in social sciences, natural sciences, and engineering grew from 58,875 to 82,266, or by almost 40 percent. We have observed, however, that for the United States to draw proportionately from these groups for STEM fields while also increasing the proportion of 24-year-olds with a first degree in STEM from 6 percent to 10 percent, we would need to roughly triple the current numbers. If we set a short-term goal of at least *doubling* the numbers in the next decade as a milestone, this would mean that the rate of change over the next 10 years would need to be 250 percent of the 1998-2007 rate (i.e., 100 percent growth rather than 40 percent growth) as we move from about 80,000 to 160,000 underrepresented minorities achieving bachelor's degrees in STEM.

To achieve this doubling, the nation must invest in our nation's students to achieve the higher level of return. *Rising Above the Gathering Storm* recommended that the United States "increase the number and proportion of U.S. citizens who earn bachelor's degrees in the physical sciences, the life sciences, engineering, and mathematics by providing 25,000 new 4-year competitive undergraduate scholarships each year to U.S. citizens attending US institutions." If one-eighth of the 80,000 additional underrepresented minority students we hope will attain a bachelor's degree in STEM over the next decade require additional financial support to persist and complete, then 10,000—or 40 percent—of these 25,000 new four-year scholarships would need to be directed to underrepresented minorities. As shown in Table 8-1, if the cost of the program were $15,000 per student per year for an institutional program plus financial support to students, the component that would support 10,000 undergraduate underrepresented minority students in STEM would cost $150 million in fiscal year 2012 for the first cohort, increasing to $600 million in 2015 and thereafter, supporting four cohorts.

Priority 2: Teacher Preparation, College Preparatory Programs, and Transition to Graduate Study

We propose also an emphasis on teacher preparation, secondary school programs that support preparation for college STEM education, and programs that support the transition from undergraduate to graduate work.

Teacher preparation may be addressed in part by providing some portion of the undergraduate support recommended above to students who make a commitment to pursue a career in K-12 science or mathematics teaching, so these are not mutually exclusive recommendations. Secondary school programs that ensure students have access to advanced courses and proper academic advising will support the goal of undergraduate persistence and completion by ensuring that matriculating freshmen are fully prepared for college study.

At the other end of the undergraduate years, the transition of underrepresented minorities to graduate work at top research universities, where they can contribute to research and leadership in our nation's science and engineering enterprise, is also critical. Equally important to the undergraduate support recommended above, we believe, is that underrepresented minorities should constitute a similar proportion of new graduate students who are supported through portable fellowships, research assistantships, or institutional grants, in order to increase their overall representation and to move greater numbers into top graduate programs. Research assistantships

TABLE 8-1 Cost Estimate for New Underrepresented Minority Student Support

	Fiscal year			
	2012	2013	2014	2015
URM Cumulative Awards	10,000	20,000	30,000	40,000
Cost for URM Cumulative Awards	$150M	$300M	$450M	$600M

Assumptions: 25,000 new students per year; 40% allocated to underrepresented minorities; 4-year scholarship; $15,000 per student for tuition/fees and institutional programs.

are particularly valuable in terms of ties to the lab group, access to research mentors and equipment, publication opportunities, and connections to professional networks.

This report resonates with the emphasis on STEM education and workforce development throughout the policy arena, including recent reports and initiatives of the Obama administration and National Science Board. It extends the previous knowledge about these issues by presenting guidance specific to the underrepresentation of minorities in STEM disciplines and careers. The report should be integral to the continued national conversations concerning the need for America to maintain a science and engineering workforce to meet its current and future needs. Finally, this is a transformative moment for the nation to seize this opportunity to not fail future generations.

BIBLIOGRAPHY

Bibliography

ACT. *The Condition of College Readiness 2009.* Iowa City, Iowa: ACT.

A Test of Leadership: Charting the Future of U.S. Higher Education. A report of the commission appointed by Secretary of Education Margaret Spellings (September 2006), p. 18-19.

Academic Competitiveness Council. 2007. *Report of the Academic Competitiveness Council.* Washington, DC: U.S. Department of Education. Available at http://www2.ed.gov/about/inits/ed/competitiveness/acc-mathscience/report.pdf.

Alesina, A., and E. La Ferrara. 2000. Participation in heterogeneous communities. *The Quarterly Journal of Economics* 115(3):847-904.

Alexander, B. B., A. C. Burda, and S. B. Millar. 1997. A community approach to learning calculus: Fostering success for undergraduate ethnic minorities in an emerging scholars program. *Journal of Women and Minorities in Science and Engineering* 3(3):145-159.

Alicia C. Dowd, Lindsey E. Malcom, and Estela Mara Bensimon. 2009. *Benchmarking the Success of Latina and Latino Students in STEM to Achieve National Graduation Goals.* University of Southern California Center for Urban Education, December 2009.

Allen, W. R. 1992. The color of success: African-American college student outcomes at predominantly White and historically Black public colleges and universities. *Harvard Educational Review* 62:45-65.

American Association for the Advancement of Science. 2001. *In Pursuit of a Diverse Science, Technology, Engineering, and Mathematics Workforce: Recommended Research Priorities to Enhance Participation by Underrepresented Minorities.* Available at http://ehrweb.aaas.org/mge/Reports/Report1/AGEP/.

American Association for Advancement of Science. 2004. *Standing Our Ground: A Guidebook for STEM Educators in the Post-Michigan Era.* Washington, DC: AAAS. Available at http://www.aaas.org/standingourground/PDFs/Standing_Our_Ground.pdf.

American Council on Education, 2005. *Increasing the Success of Minority Students in Science and Technology.* Washington, DC: ACE.

Anderson, James A. 2008. *Driving Change Through Diversity and Globalization: Transformative Leadership in the Academy.* Sterling, VA: Stylus Publishing.

Arispe, Irma. 2008. *National Mathematics Advisory Panel: Foundations for Success*. Presentation to Committee on Underrepresented Groups and the Expansion of the Science and Engineering Workforce Pipeline by White House Office of Science and Technology Policy, June 11, 2008. National Academies Keck Center, Washington, DC.

Backhus, D. 1989. Centralized intrusive advising and undergraduate retention. *NACADA Journal* 9, 27-33.

Bandura, A. 1986. The explanatory and predictive scope of self-efficacy theory. *Journal of Clinical and Social Psychology* 4(3):359-373.

Bandura, A. B. 1977. *Self-Efficacy: The Exercise of Control*. New York, NY: W. H. Freeman.

Bandura, A. B. 1985. Model causality in social learning theory. In M. J. Mahoney and A. Freeman, eds., *Cognition and Psychotherapy* (81-99). New York, NY: Plenum.

Bargh, J. and Y. Schul. 1980. On the cognitive benefits of teaching. *Journal of Educational Psychology*. 72:593-604.

Barnett, W. S., K. Jung, V. Wong, T. Cook, and C. Lamy. 2007. Effects of five state prekindergarten programs on early learning. New Brunswick, NJ: NIEER.

Basken, Paul. 2008. Education [department] blamed for not doing enough to promote grants. *Chronicle of Higher Education* (August 4). Available at http://chronicle.com/article/Education-Dept-Blamed-for-Not/1036/.

Bauer, K. W., and J. S. Bennett. 2003. Alumni perceptions used to assess the undergraduate research experience. *Journal of Higher Education* 74(2):210-230.

Bonous-Hammarth, M. 2000. Pathways to success: Affirming opportunities for science, mathematics, and engineering majors. *Journal of Negro Education* 69(1/2):92-111.

Bonsangue, M. V. and D. E. Drew. 1995. Increasing minority students' success in calculus. *New Directions for Teaching and Learning* 1995(61):23-33.

Bouffard-Bouchard, T. 1990. Influence of self-efficacy on performance in a cognitive task. *Journal of Social Psychology* 130(3):353-363.

Bowen, William G., and Derek Bok. 1998. *The Shape of the River: Long-Term Consequences of Considering Race in College and University Admissions*. Princeton, NJ: Princeton University Press.

Bowen, William G., Matthew M. Chingos, and Michael S. McPherson. 2009a. *Crossing the Finish Line: Completing College at America's Public Universities*. Princeton, NJ: Princeton University Press.

Bowen, William G., Matthew M. Chingos, and Michael S. McPherson. 2009b. Helping students finish the 4-year run," *Chronicle of Higher Education*, September 8, 2009. Available at http://chronicle.com/article/Helping-Students-Finish-the/48329.

Brayboy, Nancy, and David Begay. 2005. *Sharing the Skies: Navajo Astronomy A Cross-Cultural View*. Bluff, UT: Indigenous Education Institute.

Brown, R. P., and E. A. Day. 2006. The difference isn't black and white: Stereotype threat and the race gap on Raven's Advanced Progressive Matrices. *Journal of Applied Psychology* 91(4):979-985.

Building Engineering and Science Talent (BEST). 2004a. *A Bridge for All: Higher Education Design Principles to Broaden Participation in Science, Technology, Engineering and Mathematics*. San Diego, CA. Available at http://www.bestworkforce.org/PDFdocs/BEST_BridgeforAll_HighEdDesignPrincipals.pdf.

Building Engineering and Science Talent (BEST). 2004b. *The Talent Imperative: Meeting America's Challenge in Science and Engineering, ASAP*. San Diego, CA. Available at http://www.bestworkforce.org/PDFdocs/BEST_BridgeforAll_HighEdFINAL.pdf.

Building Engineering and Science Talent. 2004c. *What It Takes*. San Diego, CA. Available at http://www.bestworkforce.org/PDFdocs/BESTPre-K-12Rep_part1_Apr2004.pdf.

Burrelli, Joan, and Alan Rapoport. 2008. *Role of HBCUs as Baccalaureate-Origin Institutions of Black S&E Doctorate Recipients, InfoBrief*. Washington, DC: National Science Foundation (NSF 08-319).

Bush, Vannevar. *Science: The Endless Frontier*. A report to the President, July 1945.

Cervone, D., and P. K. Peaker. 1986. Anchoring, efficacy, and action: The influence of judgmental heuristics on self-efficacy judgments and behavior. *Journal of Personality and Social Psychology* 50(3):492-501.

Chang, J. C. 2003. *Women and Minorities in the Science, Mathematics, and Engineering Pipeline*. ERIC Digest. At http://www.ericdigests.org/2003-2/women.html.

Chemers, M. M., L. Hu, and B. F. Garcia. 2001. Academic self-efficacy and first-year college student performance and adjustment. *Journal of Educational Psychology* 93(1):55-64.

Chubin, Daryl E., and Wanda E. Ward. 2009. Building on the BEST principles and evidence: A framework for broadening participation. In *Broadening Participation in Undergraduate Research: Fostering Excellence and Enhancing the Impact* (M. Boyd and J. Wesermann, eds.), 13-30. Washington, DC: Council on Undergraduate Research.

Chubin, D. E., A. L. DePass, and L. Blockus, eds. 2009. Understanding Interventions That Encourage Minorities to Pursue Research Careers. Bethesda, MD: American Society for Cell Biology.

Chubin, D. E., and Malcom S. 2008. *Making a Case for Diversity in STEM Fields*. http://www.insidehighered.com/layout/set/print/viwwews/2008/10/06/chubin.

Clewell, B. C., C. C. de Cohen, L. Tsui, L. Forcier, E. Gao, N. Yung, N. Deterding, and C. West. 2006. *Final Report on the Evaluation of the National Science Foundation Louis Stokes Alliances for Minority Participation Program*. Washington, DC: The Urban Institute.

Clewell, Beatriz Chu, Clemencia Cosentino de Cohen, Nicole Deterding, and Lisa Tsui. 2006. *Revitalizing the Nation's Talent Pool in STEM*. Washington, DC: Urban Institute. Available at http://www.urban.org/publications/411301.html.

Clewell, Beatriz Chu, Clemencia Cosentino de Cohen, Nicole Deterding, and Lisa Tsui. 2010. *Capacity Building to Diversify STEM: Realizing Potential Among HBCUs*. Washington, DC: Urban Institute.

Cohen, G. L., J. Garcia, N. Apfel, and A. Master. 2006. Reducing the racial achievement gap: A social-psychological intervention. *Science* 313:1307-1310.

Cole, S., and E. Barber. 2003. *Increasing Faculty Diversity: The Occupational Choices of High-Achieving Minority Students*. Cambridge, MA: Harvard University Press.

Coleman, A. 2002. Diversity in higher education: A continuing agenda, in *Rights at Risk: Equality in an Age of Terrorism*, Report of the Citizens' Commission on Civil Rights, Washington, DC.

College Board, The. 2006. *College Board Standards for College Success: Mathematics and Statistics*. New York, NY: The College Board.

College Board, The. 2008a. *Coming to Our Senses: Education and the American Future*. New York, NY: The College Board.

College Board, The. 2008b. *Fulfilling the Commitment: Recommendations for Reforming Federal Student Aid in Brief*. The report from the Rethinking Student Aid Group.

College Board, The. 2008c. *The Report from the Rethinking Student Aid Group: Fulfilling the Commitment: Recommendations for Reforming Federal Student Aid in Brief*. New York, NY: The College Board.

College Board, The. 2008d. *Trends in Student Aid: 2008*. Trends in Higher Education Series. New York, NY: The College Board.

College Board, The. 2009. *Science: College Board Standards for College Success*. New York, NY: The College Board.

College Board, The. 2010. *The 6th Annual AP Report to the Nation*. New York, NY: The College Board.

Collier, Paul. 2000. Ethnicity politics and economic performance. *Economics and Politics* 12(3):225-245.

Consortium for Social Science Associations (CSSA). 2008. *Enhancing Diversity in Science: A Leadership Retreat on the Role of Professional Associations and Scientific Societies: A Summary Report.* Washington, DC: COSSA. Available at http://www.cossa.org/communication/diversity_workshop/Enhancing_Diversity_in_Science.pdf.

Cooper, H., B. Nye, K. Charlton, L. Lindsay, and S. Greathouse. 1996. The effects of summer vacation on achievement test scores: A narrative and meta-analytic review. *Review of Educational Research* 66:227-268.

Cooper, Laura A. 2007. Why closing the research-practice gap is critical to closing student achievement gaps. *Theory into Practice* 46(4):317-324.

Council of Graduate Schools (CGS). 2008. *Findings from the 2008 CGS International Graduate Admissions Survey. Phase III: Final Offers of Admission and Enrollment.* Washington, DC. Available at http://www.cgsnet.org/portals/0/pdf/R_intlenrl09_III.pdf.

Council of Graduate Schools. Ph.D. Completion and Attrition:Findings from Exit Surveys of Ph.D. Completers (Released September 2009). Washington, DC.

Couturier, Lara K., and Alisa F. Cunningham. 2006. *Convergence: Trends Threatening to Narrow College Opportunity in America.* Washington, DC: Institute for Higher Education Policy.

Dar-Nimrod, I., and S. J. Heine. 2006. Exposure to scientific theories affects women's math performance. *Science* 314(5798):435.

DeAngelis, Karen J., Jennifer B. Presley, and Bradford R. White. 2005. Illinois Education Research Council. Policy Report: IERC 2005-1.

Donnelly, Margarita. 1988. *Training and Recruiting Minority Teachers.* ERIC Digest Series Number EA29. At http://www.ericdigests.org/pre-9210/minority.htm.

Dowd, Alicia, and Tarek Coury. 2004. *The Effect of Loans on the Persistence and Attainment of Community College Students.* New York: Springer Netherlands. Available at http://www.springerlink.com/content/k544365421852684/fulltext.pdf.

Dynarski, S. 2004. The new merit aid. NBER Chapters in: College Choices: *The Economics of Where to Go, When to Go, and How to Pay For It,* 63-100. Cambridge, MA: National Bureau of Economic Research, Inc.

Education Trust, The. 2008. *Highlights from Trends in International Mathematics and Science Study (TIMSS) 2007.* Washington, DC: NCES.

Espenshade, T., and A. Radford. 2009. *No Longer Separate, Not Yet Equal: Race and Class in Elite College Admission and Campus Life.* Oxford, MA: Princeton University Press.

European Commission. 2003. *The Costs and Benefits of Diversity.* Kent, UK: Centre for Strategy and Evaluation Services.

Field, Kellyn. 2009. *Obama's Pell Grant Proposal Would Make 260,000 More Students Eligible.* NewsBlog: Chronicle for Higher Education. Available at http://chronicle.com/article/Obama-s-Pell-Grant-Proposal/42637.

Finn, Michael G. 2010. *Stay Rates of Foreign Doctorate Recipients from U.S. Universities, 2007.* Oak Ridge, TN: Oak Ridge Institute for Science and Education. Available at http://orise.orau.gov/sep/files/stay-rates-foreign-doctorate-recipients-2007.pdf.

Frehill, Lisa M., Nicole M. DiFabio, and Susan T. Hill. 2008. *Confronting the "New" American Dilemma—Underrepresented Minorities in Engineering: A Data-Based Look at Diversity.* Washington, DC: National Action Council for Minorities in Engineering.

Fries-Britt, S. 2000. Identity development of high ability Black collegians. *New Directions for Teaching and Learning* 82:55-65.

Fullilove, R. E., and P. U. Treisman. 1990. Mathematics achievement among African American undergraduates at the University of California, Berkeley: An evaluation of the mathematics workshop program. *Journal of Negro Education* 59(3):643-678.

Gahan-Rech, J., L. Stephens, and B. Buchalter. 1989. The effects of tutoring on college students' mathematical achievement in a mathematics laboratory. *Journal of Research and Development in Education* 22(2):18-21.

Gándara, P., and J. Maxwell-Jolly. 1999. *Priming the Pump: Strategies for Increasing the Achievement of Underrepresented Minority Undergraduates*. New York, NY: College Board.

Gansemer-Topf, Ann M., and John H. Schuh. 2005. Institutional grants: Investing in student retention and graduation *National Association of Student Financial Aid Administrators (NASFA) Journal* 35:3:5-20.

Garcia, P. (1991). Summer bridge: Improving retention rates for underprepared students. *Journal of the Freshman Year Experience* 3(2), 91-105.

George, Yolanda S., David S. Neale, Virginia Van Horne, and Shirley M. Malcom. 2001. *In Pursuit of a Diverse Science, Technology, Engineering, and Mathematics Workforce: Recommended Research Priorities to Enhance Participation by Underrepresented Minorities*. Washington, DC: American Association for the Advancement of Science and the National Science Foundation.

Gladwell's Outliers: Timing is Almost Everything. Available at http://www.businessweek.com/magazine/content/08_48/b4110110545672.html.

Gold, M. V., M. P. Deming, and K. Stone. 1992. The bridge: A summer academic enrichment program to retain African-American collegians. *Journal of the Freshman Year Experience*, 4(2):101-117.

Goldin, Claudia, and Lawrence F. Katz. 2008. *The Race Between Education and Technology*. Cambridge, MA: The Belknap Press of Harvard University Press.

Good, J., G. Halpin, and G. Halpin. 1998. The affective and academic benefits for mentors in a minority engineering program. (ERIC Document Reproduction Service No. ED429488).

Gormley, W., D. Phillips, and T. Gayer. 2008. Preschool programs can boost school readiness. *Science* 320:1723-1724.

Government Accountability Office. 2006. *Higher Education: Science, Technology, Engineering and Mathematics Trends and the Role of the Federal Government*, Statement of C. M. Ashby, Director, Education, Workforce, and Income Security Issues. Washington, DC: GAO-06-702T.

Grandy, J. 1998. Persistence in science of high-ability minority students. *The Journal of Higher Education* 69(6):589-620.

Grigg, W., M. Lauko, and D. Brockway. 2006. *The Nation's Report Card: Science 2005* (NCES 2006). U.S. Department of Education, National Center for Education Statistics. Washington, DC: NCES.

Gross, Jacob P. K., Don Hossler, and Mary Ziskin. 2007. Institutional aid and student persistence: An analysis of the effects of institutional financial aid at public four-year institutions. *National Association of Student Financial Aid Administrators (NASFAA) Journal* 37(1):28-39.

Gurin, Patricia. 1999. Expert report of Patricia Gurin prepared for *Gratz et al. v. Bollinger et al.*, No. 97-75321 (E.D. Mich.) and *Grutter et al. v. Bollinger et al.*, No. 97-75928 (E.D. Mich).

Hackett, E., J. Croissant, and B. Schneider. 1992. Industry, academe, and the values of undergraduate engineers. *Research in Higher Education* 33(3):275-95.

Hale, Frank W. 2004. *What Makes Racial Diversity Work in Higher Education*. Sterling, VA: Stylus Publishing.

Hall, D., and N. Ushomirsky. 2010. *Close the Hidden Funding Gaps in Our Schools*. Washington, DC: The Education Trust.

Harris, D. N., and T. R. Sass. 2007. *Teacher Training, Teacher Quality, and Student Achievement*. Working Papers. National Center for Analysis of Longitudinal Data in Education Research. Washington, DC: Urban Institute.

Higher Education Research Institute at UCLA (HERI). 2010. *Degrees of Success: Bachelor's Degree Completion Rates Among Initial STEM Majors, HERI Report Brief.* Available at http://www.heri.ucla.edu/nih/HERI_ResearchBrief_OL_2010_STEM.pdf (accessed February 20, 2010).

Highsmith, R. J., R. Denes, and M. M. Pierre. 1998. Mentoring Matters. (ERIC Document Reproduction Service No. ED430909).

Holton, B. E., and G. K. Horton. 1996. The Rutgers Physics Learning Center: Reforming the physics course for first-year engineering and science students. *Physics Teacher* 34(3):138-143.

Hrabowski, Freeman. 2002. Postsecondary minority student achievement: How to raise performance and close the achievement gap. *The College Board Review* 195. Princeton, NJ: The College Board.

Hunter, A.-B., S. L. Laursen, and E. Seymour. 2007. Becoming a scientist: The role of undergraduate research in students' cognitive, personal, and professional development. *Science Education* 91:36-74.

Hurtado, S., E. L. Dey, P. Gurin, and G. Gurin. 2003. The college environment, diversity, and student learning. In J.S. Smart (Ed.), *Higher education: Handbook of Theory and Research,* Vol. 18 (145-189). Amsterdam: Luwer Academic Press.

Hurtado, S., N. L. Cabrara, M. H. Lin, L. Arellano, and L. L. Espinosa. 2009. *Research in Higher Education Diversifying Science: Underrepresented Student Experiences in Structures Research Programs.* New York, NY: Springer Netherlands.

Hurtado, Sylvia, Jeffrey Milem, Alma Clayton-Pedersen, and Walter Allen. 1999. *Enacting Diverse Learning Environments: Improving the Climate for Racial/Ethnic Diversity in Higher Education, ASHE-ERIC Higher Education Report Volume 26, No. 8.* Washington, DC: The George Washington University, Graduate School of Education and Human Development.

Hussar, W. J., and T. M. Bailey. 2009. Projections of Education Statistics to 2018 (NCES 2009-062). National Center for Education Statistics, Institute of Education Sciences, U.S. Department of Education, Washington, DC.

Ingersoll, Richard M. 2008. *Core Problems: Out-of-Field Teaching Persists in Key Academic Courses and High-Poverty Schools.* Washington, DC: The Education Trust.

Institute for Higher Education Policy (IHEP). 2006. Convergence: Trends Threatening to Narrow College Opportunity in America. Washington, DC: Institute for Higher Education Policy.

Jackson, Shirley A. 2003. *The Quiet Crisis; Falling Short in Producing American Scientific and Technical Talent.* Building Engineering and Science Talent.

Johnson, M. J., J. L. Swartz, and W. E. Martin Jr. 1995. Theories for Native American career development. In F. T. L. Leong, ed., *Career Development and Vocational Behavior of Racial and Ethnic Minorities* (103-133). Mahwah, NJ: Lawrence Erlbaum Associates.

Kardash, C. M. 2000. Evaluation of undergraduate research experience: Perceptions of undergraduate interns and their faculty mentors. *Journal of Educational Psychology* 92(1):191-201.

Kelly, P. 2005. As America Becomes More Diverse: The Impact of State Higher Education Inequality National Center for Higher Education Management Systems: Boulder, CO.

Kington, Raynard. 2008. Deputy Director, National Institutes of Health, Presentation to Committee, June 11, 2008

Kochan, T., K. Bezrukova, E. Robin, A. Joshi, K. Jehn, J. Leonard, D. Levine, and D. Thomas. 2002. *The Effects of Diversity on Business Performance*: Report of the Diversity Research Network. Building Opportunities for Leadership Development Initiative, Alfred P. Sloan Foundation and the Society for Human Resource Management.

Leggon, Cheryl B., and Willie Pearson, Jr. 2008. Assessing programs to improve minority participation in STEM fields: What we know and what we need to know. In *Doctoral Education and the Faculty of the Future* (Ronald Ehrenberg and Charlotte Kuh, eds.) and citing Jackson, *The Quiet Crisis*. Ithaca, NY: Cornell University Press.

Lleras, C. 2008. Race, race concentration, and the dynamics of educational inequality across urban and suburban schools. *American Educational Research Journal* (45(4):886-913.

Locks, A. M., S. Hurtado, N. Bowman, and L. Osequera. 2008. Extending notions of campus climate and diversity to students' transition to college. *Review of Higher Education* 31(3):257-285.

Lopatto, D. 2003. The essential features of undergraduate research. *Council on Undergraduate Research Quarterly* 24:139-142.

Lopatto, D. 2004. Survey of Undergraduate Research Experiences (SURE): First findings. *Cell Biology Education* 3(4):270-277.

Lopatto, D. 2007. Undergraduate research experiences support science career decisions and active learning. *CBE–Life Sciences Education* 6(4):297-306.

Lowe, A., and M. Toney. 2001. Academic advising: Views of the givers and takers. *Journal of College Student Retention* 2(2):93-108.

Luebchow, Lindsey. 2009. *Equitable Resources in Low Income Schools: Teacher Equity and the Federal Title I Comparability Requirement*. Education Policy Program. Washington, DC: New America Foundation.

Lumina Foundation. *The Case for Improved Higher Education Access & Attainment*. Available at http://www.luminafoundation.org/our_work/ (accessed March 27, 2009)

Malcom, S., D. Chubin, and J. Jesse. 2004. *Standing Our Ground: A Guidebook for STEM Educators in the Post-Michigan Era*. Washington, DC: AAAS.

Maple, S. A., and F. K. Stage. 1991. Influences on the choice of math/science major by gender and ethnicity. *American Educational Research Journal* 28(1):37-60.

Maton, Kenneth L., and Freeman Hrabowski. 2004. Increasing the number of African American PhDs in the sciences and engineering: A strengths-based approach. American Psychologist. Washington, DC: American Psychological Association. 547-556.

Maton, Kenneth L., Freeman A. Hrabowski, Metin Ozdemir, and Harriette Wimms. 2008. Enhancing representation, retention, and achievement of minority students in higher education: A social transformation theory of change. In M. Shinn and H. H. Yoshikawa, eds., *Toward Positive Youth Development: Transforming Schools and Community Programs* (115-132). New York, NY: Oxford University Press.

McHenry, W. 1997. Mentoring as a tool for increasing minority student participation in science, mathematics, engineering, and technology undergraduate and graduate programs. *Diversity in Higher Education* 1.

Meyers, S. L., and Caroline S. V. Turner. 2000. *Faculty of Color in Academe: Bittersweet Success*. Upper Saddle River, NJ: Allyn & Bacon.

Mickelson, M. L., and M. L. Oliver. 1991. Making the short list: Black candidates and the faculty recruitment process. In P. G. Altbach and K. Lomotey (Eds.), *The Racial Crisis in American Higher Education*. Albany, NY: SUNY Press.

Milem, J. F., and K. Kakuta. 2000. The Benefits of Racial and Ethnic Diversity in *Higher Education in Minorities in Higher Education: Sixteenth Annual Status Report*. Washington, DC: American Council on Education.

Miller, D. C., A. Sen, L. B. Malley, and S. D. Burns. 2009. *Comparative Indicators of Education in the United States and Other G-8 Countries*. (NCES 2009-039). Washington, DC: U.S. Department of Education, National Center for Education Statistics.

More Diversity Among 2009 SAT Test Takers, Scores Slightly Down, *Diverse Education*, August 26, 2009. Found at http://diverseeducation.com/cache/print.php?articleId=12973.

Moreno, S. E., and C. Muller. 1999. Success and diversity: The transition through first-year calculus in the university. *American Journal of Education* 108(1):30-57.

Moust, J. C., and H. G. Schmidt. 1994. Effects of staff and students tutors on student achievement. *Higher Education* 28(4):471-82.

Murdock, T. A. 1987. It isn't just money: The effects of financial aid on student persistence. *Review of Higher Education* 11(1):75-101.

Murphy, T. J., K. L. Stafford, and P. McCreary. 1998. Subsequent course and degree paths of students in a Treisman-style workshop calculus program. *Journal of Women and Minorities in Science and Engineering* 4(4):381-396.

Myers, Jr. S., and C. Turner. 2004. The Effects of Ph.D. Supply on Minority Faculty Representation. The American Economic Review, Vol. 94, No. 2, Papers and Proceedings of the One Hundred Sixteenth Annual Meeting of the American Economic Association San Diego, CA, pp. 296-301.

Nagda, B. A., S. R. Gregerman, J. Jonides, W. von Hippel, and J. S. Lerner. 1998. Undergraduate student-faculty research partnerships affect student retention. *Review of Higher Education*, 22(1):55-72.

National Academy of Engineering. 2006. *Engineering Studies at Tribal Colleges and Universities*. Washington, DC: The National Academies Press.

National Academy of Engineering and National Research Council. 2005. *Enhancing the Community College Pathway to Engineering Careers*. Washington, DC: The National Academies Press.

National Academy of Sciences, National Academy of Engineering, and Institute of Medicine. 2007. *Rising Above the Gathering Storm: Energizing and Employing America for a Brighter Economic Future*. Washington, DC: The National Academies Press.

National Academy of Sciences, National Academy of Engineering, and Institute of Medicine. 2009. Learning Science in Informal Environments: People, Places, and Pursuits. Washington, DC: The National Academies Press.

National Action Council for Minorities in Engineering. 2008. *Confronting the "New." American Dilemma*. White Plains, NY: NACME. Available at http://www.nacme.org/user/docs/NACME%2008%20ResearchReport.pdf.

National Center for Education Statistics (NCES). 2000. *Entry and Persistence of Women and Minorities in College Science and Engineering Education* (NCES 2000-601). Washington, DC: U.S. Department of Education.

National Center for Education Statistics. 2007. *Status and Trends in the Education of Racial and Ethnic Minorities* (NCES 2010-451). Washington, DC: U.S. Department of Education. Available at http://nces.ed.gov/pubs2007/minoritytrends/.

National Center for Education Statistics. 2008.*Characteristics of Minority-Serving Institutions and Minority Undergraduates Enrolled in These Institutions*. Washington, DC: U.S. Department of Education.

National Center for Education Statistics. 2009a. Characteristics of Public, Private, and Bureau of Indian Education Elementary and Secondary School Teachers in the United States From the 2007-08 Schools Staffing Survey (NCES 2009-324). Washington, DC: National Center for Education Statistics, U.S. Department of Education.

National Center for Education Statistics. 2009b. The Nation's Report Card: Mathematics 2009. National Assessment of Educational Progress at Grades 4 and 8 (NCES 2010-451). Washington, DC: National Center for Education Statistics, U.S. Department of Education.

National Center for Education Statistics, 2010. *Status and Trends in the Education of Racial and Ethnic Minorities* (NCES 2010-015), July 2010. http://nces.ed.gov/pubs2007/minoritytrends/ (accessed July 15, 2010).

National Math and Science Initiative. Tackling the STEM Crisis: Five Steps Your State Can Take to Improve the Quality and Quantity of Its K-12 Math and Science Teachers. e. Available at http://www.nctq.org/p/docs/nctq_nmsi_stem_initiative.pdf.

National Research Council. 1999. *Improving Student Learning: A Strategic Plan for Education Research and Its Utilization.* Washington, DC: National Academy Press.

National Research Council. 2003. *BIO2010: Transforming Undergraduate Education for Future Research Biologists.* Washington, DC: The National Academies Press.

National Research Council. 2005a. *Assessment of the NIH Minority and Research and Training Programs.* Washington, DC: The National Academies Press.

National Research Council. 2005b. *Enhancing the Community College Pathway to Engineering Careers.* Washington, DC: The National Academies Press.

National Research Council. 2007a. *Beyond Bias and Barriers: Fulfilling the Potential of Women in Academic Science and Engineering.* Washington, DC: The National Academies Press.

National Research Council. 2007b. *Ready, Set, Science! Putting Research to Work in K-8 Science Classrooms.* Washington, DC: The National Academies Press.

National Research Council. 2007c. *Understanding Interventions That Encourage Minorities to Pursue Research Careers.* Washington, DC: The National Academies Press. Available at http://books.nap.edu/catalog.php?record_id=12022.

National Research Council. 2008. *Science Professionals: Master's Education for a Competitive World.* Washington, DC: The National Academies Press.

National Research Council. 2009. *Partnerships for Emerging Research Institutions: Report of a Workshop.* Washington, DC: The National Academies Press.

National Science and Technology Council. 2000. *Ensuring a Strong U.S. Scientific, Technical, and Engineering Workforce in the 21st Century.* Washington, DC: Office of Science and Technology Policy.

National Science Board Commission on Precollege Education in Mathematics, Science and Technology. 1983. *Educating Americans for the 21st Century.* National Science Foundation.

National Science Board. 2007. *National Action Plan for Addressing the Critical Needs of the U.S. Science, Technology, Engineering, and Mathematics Education System.* Arlington, VA: National Science Foundation. Available at http://www.nsf.gov/nsb/publications/pub_summ.jsp?ods_key=nsb07114.

National Science Board. 2008. *Science and Engineering Indicators.* 2 vols. Arlington, VA: National Science Foundation.

National Science Board. 2009. *Science and Engineering Indicators.* 2 vols. Arlington, VA: National Science Foundation.

National Science Board. 2010. *Science and Engineering Indicators.* 2 vols. Arlington, VA: National Science Foundation.

National Science Foundation. 1998. *The Federal Role in Science and Engineering Graduate and Postdoctoral Education.* Arlington, VA: National Science Foundation.

National Science Foundation. 1999. *Retention of the Best Science and Engineering Graduates in Science.* Arlington, VA: National Science Foundation.

National Science Foundation. 2000. *Land of Plenty: Diversity as America's Competitive Edge in Science, Engineering and Technology.* Arlington, VA: National Science Foundation. Available at http://www.nsf.gov/pubs/2000/cawmset0409/cawmset_0409.pdf.

National Science Foundation. 2001. *Human Resources Contributions to U.S. Science and Engineering from China.* (SRS InfoBrief 2001). Arlington, VA: National Science Foundation.

National Science Foundation. 2008. *Doctorate Recipients from U.S. Universities, Summary Report 2007-08* (NSF 10-309). Arlington, VA: National Science Foundation.

National Science Foundation. 2009. *Women, Minorities, and Persons with Disabilities in Science and Engineering.* Arlington, VA: National Science Foundation. Available at http://www.nsf.gov/statistics/wmpd.National Center for Education Statistics, 2010. *Status and Trends in the Education of Racial and Ethnic Minorities* (NCES 2010-015), July 2010. http://nces.ed.gov/pubs2007/minoritytrends/ (accessed July 15, 2010).

NCTQ and NMSI. 2009. *Tackling the STEM Crisis: Five Steps Your State Can Take to Improve the Quality and Quantity of Its K–12 Math and Science Teachers.* Available at http://www.nctq.org/p/docs/nctq_nmsi_stem_initiative.pdf.

Nelson, Donna J. 2009. *A National Analysis of Minorities in Science and Engineering Faculties at Research Universities.* Norman, OK: University of Oklahoma. Available at: http://chem.ou.edu/~djn/diversity/Faculty_Tables_FY07/FinalReport07.html.

Nettles, M. T., ed. 1988. *Toward Black Undergraduate Student Equality in American Higher Education.* New York, NY: Greenwood Press.

Newfield, Christopher. 2008. *Unmaking the Public University: The Forty Year Assault on the Middle Class.* Cambridge, MA: Harvard University Press.

Novels, A. N., and S. C. Ender. 1988. The impact of developmental advising for high-achieving minority students. *NACADA Journal* 8(2):23-26.

Obama, Barack. 2009. Address to Joint Session of Congress, February 24, 2009. Available at http://www.whitehouse.gov/the_press_office/remarks-of-president-barack-obama-address-to-joint-session-of-congress/ (accessed September 4, 2009).

Olson, S., and A. P. Fagen, eds. 2007. *Understanding Interventions That Encourage Minorities to Pursue Research Careers: Summary of a Workshop.* Washington, DC: The National Academies Press.

Orszag, Peter, and John Holdren. 2009. *Memorandum for Heads of Executive Departments and Agencies: Science and Technology Priorities for FY 2011.* Washington, DC: White House. Available at househttp://www.whitehouse.gov/sites/default/files/Final%20Signed%20OMB-OSTP%20Memo%20-%20ST%20Priorities.pdf.

Oseguera, L., S. Hurtado, N. Denson, O. Cerna, and V. Saenz. 2006. The characteristics and experiences of minority freshman committed to biomedical and behavioral science research careers. *Journal of Women and Minorities in Science and Engineering* 12(2/3):155-177.

Page, Scott. 2007. *The Difference: How the Power of Diversity Creates Better groups, Firms, Schools, and Societies.* Woodstock, Oxfordshire, U.K.: Princeton University Press.

Pascarella, E. T. and J. R. Staver. 1985. The influence of on-campus work on science career choice during college: A causal modeling approach. *Review of Higher Education* 8(3):229-245.

Pascarella, E. T., and P. T. Terenzini. 1991. *How college affects students.* San Francisco, CA: Jossey-Bass Publishers.

Pearson, Willie, Jr. 2005. *Beyond Small Numbers: Voices of African American PhD Chemists.* Stamford, CT: Jai Press.

Pearson, Willie, Jr., and Diane Martin. 2005. *Broadening Participation Through a Comprehensive, Integrated System.* Arlington, VA: National Science Foundation.

Penick, B. E., and C. Morning. 1983. The retention of minority engineering students. Report on the 1981-82 NACME retention research program. (ERIC Document Reproduction Service No. ED247325).

Perna, L. W., S .L. Fries-Britt, D. S. Gerald, H. T. Rowan-Kenyon, and J. Milem. 2008. Underrepresentation in the academy: A study of race equity in three southern states. *Journal of the Professoriate* 3(1):5-28.

Peske, H., and K. Haycock. 2006. *Teaching Inequality: How Poor and Minority Students Are Shortchanged on Teacher Quality.* Washington, DC: The Education Trust.

Planty, M., W. Hussar, T. Synder, G. Kena, A. KewalRamani, J. Kemp, K. Bianco, R. Dinkes, K. Ferguson, A. Linvingston, and T. Nachazel. 2009. *The Condition of Education 2009* (NCES 2009-081). Washington, DC: National Center for Education Statistics Institute of Education Sciences, U.S. Department of Education.

Plummer, Janelle L. 2009. *More Diversity Among 2009 SAT Test-Takers, Scores Slightly Down.* Fairfax, VA: CMA publication. Available at http://diverseeducation.com/cache/print.php?articleId=12973.

Redmond, S. P. 1990. Mentoring and cultural diversity in academic settings. *American Behavioral Scientist* 34(2):188-200.

Rubenstein, E. S., and the Staff of NPI. 2008. *Cost of Diversity: The Economic Costs of Racial and Cultural Diversity*, Issue Number 803. Augusta, GA: National Policy Institute.

Rueckert, L. 2002. Report from CUR 2002: Assessment of research. *Council on Undergraduate Research Quarterly* 23:10-11.

Russell, S. H., M. P. Hancock, and J. McCullough. 2007. Benefits of undergraduate research experiences. *Science* 316(5824):548-549.

Ryu, Mikyung. 2008. *Minorities in Higher Education*. Washington, DC: American Council on Education.

Santos, S. J. and E. T. Reigadas. 2002. Latinos in higher education: An evaluation of a university faculty mentoring program. *Journal of Hispanic Higher Education* 1:40-50.

Schunk, D. H. 1981. Modeling and attributional effects on children's achievement: A self-efficacy analysis. *Journal of Educational Psychology* 73(1):93-105.

Schwitzer, A. M., and R. Thomas. 1998. Implementation, utilization, and outcomes of a minority freshman peer mentor program at a predominantly white university. *Journal of the Freshman Year Experience and Students in Transition* 10(1):31-50.

Seymour, Elaine, and Nancy M. Hewitt. 1997. *Talking About Leaving: Why Undergraduates Leave the Sciences*. Boulder, CO: Westview.

Shapley, D., and R. Roy. 1985. *Lost at the Frontier: U.S. Science and Technology Policy Adrift*. Philadelphia, PA: ISI Press.

Slaughter, John Brooks. 2008. The *"New" American Dilemma, An Open Letter from Dr. John Brooks Slaughter, in NACME, Confronting the "New" American Dilemma: Underrepresented Minorities in Engineering: A Data-based Look at Diversity*. Washington, DC: NACME.

Smith, D. 1997. *Diversity Works: The Emerging Picture of How Students Benefit*. Washington, DC: Association of American Colleges and Universities.

Spencer, S. J., C. M. Steele, and D. M. Quinn. 1999. Stereotype threat and women's math performance. *Journal of Experimental Social Psychology* 35(1):4-28.

Springer, L., M. E. Stanne, and S. S. Donovan. 1999. Effects of small-group learning on undergraduates in science, mathematics, engineering, and technology: A meta-analysis. *Review of Educational Research* 69(1):21-51.

Steele, C. M. 1992. April. Race and the schooling of Black Americans. *Atlantic Monthly* 1-17.

Steele, C. M., and J. Aronson. 1995. Stereotype threat and the intellectual test performance of African Americans. *Journal of Personality and Social Psychology* 69(5):797-811.

Steele, C. M., S. J. Spencer, and J. Aronson. 2002. Contending with group image: The psychology of stereotype and social identity threat. *Advances in Experimental Social Psychology* 24:379-440.

Summers, Michael. 2008. *Remarks to High-Achieving Minority High School Students*. University of Maryland Baltimore County.

Summers, M. F. and F. Hrabowski. 2006. Preparing minority scientists and engineers. *Science* 311:1870-1871.

Swail, W. S., K. E., Redd, and L. W. Perna. 2003. *Retaining Minority Students in Higher Education*. ASHE-ERIC Higher Education Report. Los Angeles, CA: Jossey-Bass.

Thile, E. L., and G. E. Matt. 1995. The ethnic mentor undergraduate program: A brief description and preliminary findings. *Journal of Multicultural Counseling and Development* 23(2):116-126.

Treisman, U. 1992. Studying students studying calculus: A look at the lives of minority mathematics students in college. *College Mathematics Journal* 23(5):362-372.

Trippi, J. and H. E. Cheatham. 1991. Counseling effects on African American college students graduation. *Journal of College Student Development* 32(4):342-349.

Turner, Sarah and John Bound. 2003. Closing the Gap or Widening the Divide: The Effects of the G.I. Bill and World War II on the Educational Outcome of Black Americans. *Journal of Economic History*. Available at http://en.wikipedia.org/w/index.php?title=African_Americans_and_the_G.I._Bill&printabl.

U.S. Census Bureau, Population estimates, http://www.census.gov/popest/national/asrh/NC-EST2007-asrh.html (accessed March 27, 2009).

U.S. Department of Education. 2010. A Blueprint for Reform: The Reauthorization of the Elementary and Secondary Education Act. U.S. Department of Education, March 2010.

U.S. Department of Health and Human Services Administration for Children and Families January 2010 Head Start Impact Study.

U.S. Department of Labor, Bureau of Labor Statistics, Current Employment Statistics, Employment Situation, Table A-1, Employment status of the civilian population by sex and age, http://www.bls.gov/news.release/empsit.t01.htm (accessed June 16, 2009).

Walters, N. B. 1997. Retaining aspiring scholars: Recruitment and retention of students of color in graduate and professional science degree programs. ASHE Annual Meeting Paper. ERIC Document Reproduction Service No. ED415816. Available at http://www.eric.ed.gov/PDFS/ED415816.pdf.

Ware, N. C., and V. E. Lee. 1988. Sex differences in choice of college science majors. *American Educational Research Journal* 25(4):593-614.

Warner, Isiah. 2008. "A Tale of Three Chemists," Presentation to Study Committee, Third Committee Meeting, October 22, 2008.

Williams, D. A., J. B. Berger, and S. A. McClendon. 2005. Toward a Model of Inclusive Excellence and Change in Postsecondary Institutions. Washington, DC: Association of American Colleges and Universities.

Winton, Pam, and V. Buysse. (eds.) 2005. *Early Developments Magazine* 9:1. Chapel Hill, NC: University of North Carolina Press.

Zimmerman, B. J. (1989). A social-cognitive view of self-regulated academic learning. *Journal of Educational Psychology* 81(3):329-339.

APPENDIXES

Appendix A

Charge to the Study Committee

An ad hoc committee, under the aegis of the Committee on Science, Engineering, and Public Policy (COSEPUP), will explore the role of diversity in the science, technology, engineering, and mathematics workforce and its value in keeping America innovative and competitive. The study will analyze the rate of change and the challenges the nation currently faces in developing a strong and diverse workforce. It will identify best practices and the characteristics of these practices that make them effective and sustainable.

The committee will respond to the following questions:

(1) What are the key social and institutional factors that shape decisions of minority students to commit to education and careers in the science, technology, engineering, and math fields? What programs have successfully influenced these factors to yield improved results?

(2) What are the specific barriers preventing greater minority student participation in the science, technology, engineering, and math fields? What programs have successfully minimized these barriers?

(3) What are the primary focus points for policy intervention to increase the recruitment and retention of underrepresented minorities in America's workforce in the future? Which programs have successfully implemented policies to improve recruitment and retention? Are they "pull" or "push" strategies?" Overall, how effective have they been? By what criteria should they be judged?

(4) What programs are under way to increase diversity in the science, technology, engineering, and math fields? Which programs have been shown

to be effective? Do they differ by gender within minority group? What factors make them more effective? How can they be expanded and improved in a sustainable way?

(5) What is the role of minority-serving institutions in the diversification of America's workforce in these fields? How can that role be supported and strengthened?

(6) How can the public and private sectors better assist minority students in their efforts to join America's workforce in these fields?

(7) What should be the implementation strategy? The committee should develop a prioritized list of policy and funding action items (e.g., tax credits) with milestones and cost estimates that will lead to a science and engineering workforce that mirrors the nation's diverse population.

Appendix B

U.S. Senate Letter to the
National Academy of Sciences

EDWARD M. KENNEDY
MASSACHUSETTS

United States Senate

WASHINGTON, DC 20510–2101

November 17, 2006

Ralph J. Cicerone, PhD
President
National Academy of Sciences
500 5th Street, NW
Washington, DC 20001

Dear Dr. Cicerone:

We understand that the National Academy of Sciences has begun to examine the issue of diversity in the science, technology, engineering, and math workforce and its role in keeping America innovative and competitive. We are writing to urge you to conduct a formal study on this critical issue.

The National Academy's recent report, *Rising Above the Gathering Storm: Energizing and Employing America for a Brighter Economic Future,* identified key policy recommendations for maintaining American competitiveness in the years to come. The report ranked the improvement of education in science, technology, engineering, and math, at the top of the list, and rightly recognized the need to ensure that low income and minority students have equal access to the highest quality education in these disciplines. By exploring the challenges we currently face and the steps we can take to overcome them, the Academy can have a critical role in developing a strong and diverse workforce and in ensuring that each and every individual in America has the opportunity to contribute to the 21st century economy.

Currently, African-Americans represent only 4% of the science and engineering workforce, and Hispanics are similarly underrepresented. Americans of color are entering postsecondary studies in science, technology, engineering, and math at higher rates than white students, but significantly fewer African-American and Hispanic students are graduating with degrees in these fields. Only 63% of such students will continue their study in such fields to earn a bachelor's degree.

By 2050, the Census Bureau predicts that 50% of the U.S. population will be Hispanic, African-American, and Asian. The current science and engineering workforce, however, is nearly 82 percent white and more than 75 percent male. If this workforce is to expand as predicted, and if we are to continue to be innovative and competitive, we cannot accept this decreasing percentage of the population in fields that are so vital to the future of the American economy.

We ask that you explore challenges and barriers to diversity at every level of education and employment in the fields of science, technology, engineering, and math and make recommendations for ways the federal government can better ensure a robust, diverse workforce in the years to come. Specifically, we ask that you examine the following:

- What are the key social and institutional factors that shape the decisions of minority students to commit to education and careers in the science, technology, engineering, and math fields?
- What are the specific barriers preventing greater minority student participation in the science, technology, engineering, and math fields?
- What are the primary focus points for policy intervention to increase the recruitment and retention of underrepresented minorities in America's workforce in the future?
- What programs are underway to increase diversity in the science, technology, engineering, and math fields? Which programs have been shown to be effective? How can they be expanded and improved?
- What is the role of minority-serving institutions in the diversification of America's workforce in these fields? How can that role be supported and strengthened?
- How can the public and private sectors better assist minority students in their efforts to join America's workforce in these fields?

The National Academy's insight on these issues will provide us with needed guidance on how we can work most affectively to develop a strong and diverse workforce in these fields that is equipped for success in the global economy of the 21st Century and beyond. As you know, our diversity is one of our greatest strengths, but we cannot fully utilize that strength unless we ensure that all Americans are given the educational and career opportunities they need and deserve. We thank the National Academy for its commitment to this fundamental issue and for its consideration of our request to conduct further study in this area as soon as possible.

Sincerely,

_____
Edward M. Kennedy

_____
Barbara A. Mikulski

_____
Patty Murray

_____
Hillary Rodham Clinton

cc: Dr. Deborah Stine, PhD, Associate Director, Committee on Science, Engineering, and Public Policy

Appendix C

Committee Member Biographies

FREEMAN A. HRABOWSKI, III (*Chair*), has served as president of The University of Maryland, Baltimore County (UMBC) since May 1992. He serves as a consultant to the National Science Foundation and the National Institutes of Health. He also sits on several corporate and civic boards, such as the Carnegie Foundation for the Advancement of Teaching, Constellation Energy Group, the France-Merrick Foundation, Marguerite Casey Foundation (Chair), McCormick & Company, Inc., Mercantile Safe Deposit & Trust Company, and the Urban Institute. He has coauthored two books, *Beating the Odds* and *Overcoming the Odds* (Oxford University Press), focusing on parenting and high-achieving African American males and females in science. Both books are used by universities, school systems, and community groups around the country. Born in 1950 in Birmingham, Alabama, Dr. Hrabowski graduated at 19 from Hampton Institute with highest honors in mathematics. At the University of Illinois at Urbana-Champaign, he received his MA in mathematics and four years later his PhD in higher education administration/statistics at age 24. He has served on the National Academies' Government-University-Industry Research Roundtable and the Oversight Committee for the NRC's Assessment for NIH Minority Research/Training Programs. He also recently provided testimony for a hearing on women in academic science and engineering hosted by the Research and Science Education Subcommittee of the House Committee on Science and Technology.

JAMES H. AMMONS is the president of Florida Agricultural and Mechanical University (FAMU). A native of Winter Haven, Florida, Ammons earned his baccalaureate degree at FAMU, graduating in 1974 with a degree in political science. He then enrolled at Florida State University, earning a master's degree in public administration in 1975 and a doctorate in government in 1977. Ammons began his academic career at the University of Central Florida, where he served as an assistant professor of public administration from 1977 to 1983 before returning to FAMU as a professor of political science. Over the course of his 17-year tenure at FAMU, Ammons has risen steadily through the administrative ranks, serving as assistant vice president for academic affairs from 1984 to 1989. From 1987 to 1988, he also served as a faculty program consultant to the Board of Regents, leading a comprehensive review of the Florida system's nine political science programs. In 1989, Ammons was promoted to associate vice president for academic affairs and director of Title III programs, a post he held until being named FAMU's chief academic officer in 1995. During his tenure as provost and vice president for academic affairs, the campus has witnessed unprecedented growth in student enrollment, freshman SAT/ACT scores, retention and graduation rates, and academic program offerings. Active in professional and civic organizations, Ammons has received numerous awards and honors. He was named an American Council on Education Fellow and a CIGNA Foundation Fellow in 1986-1987, a Booth Ferris Fellow at the University of Wisconsin-Madison in 1993, and a Nissan-Educational Testing Service Fellow in 2000. At FAMU, he has received the 1987 Distinguished Alumni Award and the 1999 Millennium Award. Ammons was recently elected to the Commission on Colleges of the Southern Association of Colleges and Schools and has chaired numerous SACS accreditation review committees, including the 1999 SACS review of NCCU. He currently chairs the Editorial Board of the University Press of Florida and serves on the American Association of State Colleges and Universities' Task Force on the Professional Development of Teachers. He also has served on the boards of directors of the Greater Tallahassee Chamber of Commerce and the Tallahassee Marine Institute, as well as on the Promotion Review Board of the Florida Highway Patrol.

SANDRA BEGAY-CAMPBELL is a principal member of the technical staff at Sandia National Laboratories. Ms. Begay-Campbell leads Sandia's technical efforts in the Renewable Energy Program to assist tribes with renewable energy development. She also serves as a member of the National Science Foundation's Committee on Equal Opportunities in Science and Engineering. A member of the Navajo nation, she combines her cultural values with the technical environment. Begay-Campbell is the former executive director of the American Indian Science and Engineering Society (AISES), a nonprofit organization whose mission is to increase the number of American Indian scientists and engineers. She received a BA in civil engineering from the University of New Mexico and worked at Lawrence Livermore National Laboratories before she earned an MA in structural engineering from Stanford. She subsequently worked at Los Alamos National Laboratory before joining Sandia. Begay-Campbell has served on two committees for the National Academy of Engineering, the Committee on Diversity of the Engineering Workforce, and the Committee on Engineering Studies at Tribal Colleges.

BEATRIZ CHU CLEWELL, director of the Program for Evaluation and Equity Research (PEER) and principal research associate in the Urban Institute's Education Policy Center, is a leading expert on breaking barriers to move more women and underrepresented minorities into the science and technology workforce. Her recent journal article, "Taking Stock: Where We've Been, Where We Are, Where We're Going," traces women's progress in science, technology, engineering, and mathematics (STEM) over the past decade. She is also an author of a 2005 review of mathematics and science curricula and professional development models for middle and high school levels proven effective in increasing student achievement. An author of *Breaking the Barriers: Helping Female and Minority Students Succeed in Mathematics and Science*, she explored the theoretical and empirical foundations of intervention programs to increase the success of women and underrepresented minorities in science and mathematics. In 2007 she published *Effective Schools in Poor Neighborhoods: Defying Demographics, Achieving Success*. Dr. Clewell received a BA in English literature and a PhD in educational policy from Florida State University. She was a senior research scientist at Educational Testing Service for 12 years before joining the Urban Institute and, on leave from that organization, served at the National Science Foundation as Executive Director of a bipartisan commission on the status of women, minorities, and persons with disabilities in STEM (CAWMSET). She served on the Committee on Science Education K-12 and the Committee on NASA Education Program Outcomes Study for the NRC. Dr. Clewell has been the principal investigator for several formal evaluations of major NSF intervention programs to increase the participation of women and minorities in STEM, including the Louis Stokes Alliance for Minority Participation (LSAMP), the Program for Women and Girls, and HBCU-UP.

NANCY S. GRASMICK is Maryland's first female State Superintendent of Schools. She has served in that post since 1991. Dr. Grasmick's career in education began as a teacher of deaf children at the William S. Baer School in Baltimore City. She subsequently served as a classroom and resource teacher, principal, supervisor, assistant superintendent, and associate superintendent in the Baltimore County Public Schools. In 1989, she was appointed Special Secretary for Children, Youth, and Families and, in 1991, the State Board of Education appointed her State Superintendent of Schools. Dr. Grasmick holds a PhD from the Johns Hopkins University, an MS from Gallaudet University, and a BS from Towson University. Dr. Grasmick has been a teacher and an administrator, and, most importantly, a child advocate. Her numerous board and commission appointments include the President's Commission on Excellence in Special Education, the U.S. Army War College Board of Visitors, the Towson University Board of Visitors, State Planning Committee for Higher Education, and the Maryland Business Roundtable for Education. Dr. Grasmick has received numerous awards for her visionary leadership, including the Harold W. McGraw, Jr. Prize in Education.

CARLOS G. GUTIERREZ is professor of chemistry and biochemistry at California State University, Los Angeles. He received a PhD in synthetic organic chemistry from the University of California, Davis, in 1975. Dr. Gutierrez was a visiting scholar at the Department of Chemistry at the University of California, Berkeley. At California State University, Los Angeles, he has served as director of the Access to Research Careers program since 1978, the Minority Student Training for Biomedical Research program since 1992, the Los Angeles Bridges to the Future program from 1993 to 1997, and the Los Angeles Biomedical Sciences program from 1980 to 1983. Dr. Gutierrez has been a member of the National Advisory General Medical Sciences Council of the NIH since 1995. He has served as a member of the National Research Council's Advisory Committee of the Office of Scientific and Engineering Personnel and Board on Higher Education and Workforce. Additionally, he has served as vice chair of the NRC's Committee on the Feasibility of a National Scholars Program and was chair of the Oversight Committee for the Assessment for NIH Minority Research/ Training Programs: Phase 3 for PGA. He has also served on the California State University Systemwide Task Force on the Status of Women Faculty and Students in the Sciences.

EVELYNN M. HAMMONDS is the dean of Harvard College, starting in 2008, and Barbara Gutmann Rosenkrantz Professor of the History of Science and of African American Studies. She was the Senior Vice Provost for Faculty Development and Diversity at Harvard University. She has published articles on the history of disease, race and science, African American feminism, African American women and the epidemic of HIV/AIDS, and analyses of gender and race in science and medicine. She is also the author of the article "Gendering the Epidemic: Feminism and the Epidemic of HIV/ AIDS in the United States, 1981-1999," which appears in *Science, Medi-*

cine, and Technology in the 20th Century: What Difference Has Feminism Made? (2000). Dr. Hammonds' current work focuses on the intersection of scientific, medical, and sociopolitical concepts of race in the United States. She is completing a history of biological, medical, and anthropological uses of racial concepts entitled *The Logic of Difference: A History of Race in Science and Medicine in the United States, 1850–1990.* She is also completing the *MIT Reader on Race and Gender in Science,* coedited with Rebecca Herzig and Abigail Bass. Dr. Hammonds was named a Sigma Xi Distinguished Lecturer (2003–2005) by Sigma Xi, the Scientific Research Society. She has been a visiting scholar at the Max Planck Institute for the History of Science in Berlin and a fellow in the School of Social Science at the Institute for Advanced Study in Princeton. She earned a PhD in the history of science from Harvard University, an MS in physics from the Massachusetts Institute of Technology, a BEE in electrical engineering from the Georgia Institute of Technology, and a BS in physics from Spelman College. She taught at the Massachusetts Institute of Technology before coming to Harvard. While at MIT she was the founding director of the MIT Center for the Study of Diversity in Science, Technology, and Medicine. Dr. Hammonds has been a Visiting Professor at UCLA and Hampshire College.

WESLEY L. HARRIS is head of the MIT Department of Aeronautics and Astronautics, where he is the Charles Stark Draper Professor of Aeronautics. He also serves as vice chair of the National Science Foundation Committee on Equal Opportunities in Science and Education, which has been tasked by Congress to address problems of growth and diversity in science and engineering. He is a former NASA associate administrator for aeronautics, responsible for all aeronautics programs, facilities, and personnel (1993-1995). From 1990 to 1993 he was the University of Tennessee Space Institute's vice president and chief administrative officer. From 1985 to 1990, he served as Dean of the School of Engineering and Professor of Mechanical Engineering at the University of Connecticut. From 1972 to 1985, he held faculty and administrative positions at MIT, including professor of aeronautics and astronautics. His academic research with unsteady aerodynamics, aero acoustics, rarefied gas dynamics, sustainment of capital assets, and chaos in sickle cell disease have made seminal contributions in these fields. In academe, Dr. Harris has worked with industry and governments to design and build joint industry-government-university research and development programs, centers, and institutes. An elected fellow of the AIAA and of the AHS, Dr. Harris was recognized for personal engineering achievements, engineering education, management, and advancing cultural diversity. He has been recognized by election to membership in the National Academy of Engineering, the Cosmos Club, and the Confrérie des Chevaliers du Tastevin. He is a member of the following NRC committees: Committee on Assessing Corrosion Education (Chair), Committee on Engineering Education, Committee on Systems Engineering: A Retrospective Review and Benefits for Future Air Force Systems Acquisition, Air Force Studies Board, and Division Committee on Engineering and Physical Sciences.

SYLVIA HURTADO is professor and director of the Higher Education Research Institute at UCLA in the Graduate School of Education and Information Sciences. Just prior to coming to UCLA, she served as director of the Center for the Study of Higher and Postsecondary Education at the University of Michigan. Dr. Hurtado has published numerous articles and books related to her primary interest in student educational outcomes, campus climates, college impact on student development, and diversity in higher education. She has served on numerous editorial boards for journals in education and served on the boards for the American Association of Higher Education (AAHE) and the Higher Learning Commission and is past president of the Association for the Study of Higher Education (ASHE). Black Issues in Higher Education named her among the top 15 influential faculty whose work has had an impact on the academy. She obtained her PhD in education from UCLA, MEd from Harvard Graduate School of Education, and AB from Princeton University in sociology. Dr. Hurtado has coordinated several national research projects, including a U.S. Department of Education-sponsored project on how colleges are preparing students to achieve the cognitive, social, and democratic skills to participate in a diverse democracy. She is launching a National Institutes of Health project on the preparation of underrepresented students for biomedical and behavioral science research careers. She has also studied assessment, reform, and innovation in undergraduate education on a project through the National Center for Postsecondary Improvement.

JAMES S. JACKSON is director of the Institute for Social Research (ISR) at the University of Michigan. He is the Daniel Katz Distinguished University Professor of Psychology and directs the ISR Research Center for Group Dynamics and the Program for Research on Black Americans, which he helped to establish in 1976 at the ISR. In addition, Jackson is a professor of health behavior and health education at the U-M School of Public Health and directs the U-M Center for Afro-American and African Studies. In 1980, he directed the National Survey of Black Americans, the first survey of a nationally representative sample of Black Americans. In 2002, Jackson was elected a member of the Institute of Medicine of the National Academies of Science. Jackson is immediate past chair of the Section on Social, Economic, and Political Sciences (K) of the American Association for the Advancement of Science. He is a former chair of the Section on Social and Behavioral Sciences and of the Task Force on Minority Issues of the Gerontological

Society of America, the Committee on International Relations, and the Association for the Advancement of Psychology, American Psychological Association. He was a recipient of a Fogarty Senior Postdoctoral International Fellowship, 1993-1994, for study in France and Western Europe. He is former national president of the Black Students Psychological Association and the Association of Black Psychologists. Jackson received a BS degree in psychology from Michigan State University in 1966, an MA in psychology from the University of Toledo in 1970, and a PhD in social psychology from Wayne State University in 1972. He currently serves on the U.S. National Committee for the International Union of Psychological Science, Committee on Health Research and the Privacy of Health Information: The HIPAA Privacy Rule, and the Committee on International Collaborations in Social and Behavioral Research for the National Academies.

SHIRLEY MATHIS McBAY is the president of Quality Education for Minorities (QEM) Network. Dr. McBay earned the BA in chemistry from Paine College (1954) and an MS in chemistry from Atlanta University in 1957. In mathematics, she earned an MS from Atlanta University (1958), and a PhD at the University of Georgia (1966). In 1972 she was director of the Division of Natural Sciences at Spelman College. After administering National Science Foundation programs for five years, Dr. McBay became Dean for Student Affairs at the Massachusetts Institute of Technology. In 1990 she left this position to become president of the QEM Network, a nonprofit educational organization that was the successor to the MIF-based QEM project. QEM is dedicated to improving education for minorities at all educational levels. She served on the Advisory Board for the National Science Resources Center and on the Maryland Public Broadcasting Commission.

DIANA NATALICIO is president of the University of Texas at El Paso (UTEP). Prior to her appointment as president in 1988, Natalicio served as UTEP's vice president for academic affairs, dean of the College of Liberal Arts, and chair of the Modern Languages Department. She has written numerous books, monographs, and articles in the field of applied linguistics. Dr. Natalicio has served on the Presidential Advisory Commission on Educational Excellence for Hispanic Americans, National Action Council for Minorities in Engineering, the Fund for Improvement of Post-Secondary Education, the National Science Board, and the board of directors for the Fogarty International Center at the NIH. She was also the chair of the HCBU/MSI Consortium on Environmental Technology. She is the recipient of the 1997 Harold W. McGraw, Jr. Prize in Education, the 1991 Torch of Liberty Award from the Anti-Defamation League of B'nai B'rith, the 1990 Conquistador Award from the City of El Paso, and the 2006 Distinguished Alumnus Award from U. Texas-Austin. She has been named to both the

El Paso Women's Hall of Fame and Texas Women's Hall of Fame. She completed her undergraduate studies in Spanish at St. Louis University and earned a master's degree in Portuguese and a doctorate in linguistics from the University of Texas at Austin. She chaired on the Committee on Partnerships for Emerging Research Institutions for the National Research Council.

JOHN C. NEMETH is vice president of Oak Ridge Associated Universities (ORAU). In this role, he is responsible for developing partnerships with government and the private sector on behalf of the 117-member science and technology based consortium of colleges and universities nationwide. ORAU invests nearly $2.5 million annually in activities that benefit the students and faculty of its membership. He also manages an ORAU National Security Experts Team, composed of academic experts, to assist elements of the U.S. Department of Homeland Security in the event of a weapon of mass destruction and its effect on the United States. Additionally, he has been heavily involved with ORAU's Historically Black College and University/Minority Educational Institutions Council, which builds relationships between minority-serving institutions and research-intensive universities and federal labs. He also participated in Oak Ridge National Laboratory/ORAU HBCU/MEI Faculty Summer Outreach Program, which provides opportunities for faculty at HCBU/MEIs to build collaborations with Oak Ridge staff. Dr. Nemeth was head of the Environmental Science and Technology Program of the Georgia Tech Research Institute. Prior to this position, he was chief of the Environmental Health and Safety Division and was also program manager of Hazardous and Industrial Waste. Before joining Georgia Tech, Dr. Nemeth was director of Environmental Sciences-Eastern District and Senior Waste Management Specialist for CH2M HILL. Earlier, Dr. Nemeth was chief scientist and corporate environmental sciences consultant for Law Engineering Testing Company (Law Environmental). As senior ecologist for Coastal Zone Resources Corporation, he managed numerous environmental assessment projects. His project experience, national in scope, spans both the public and private sectors, including the complete spectrum of hazardous, industrial, and domestic waste management, environmental services and assessment work, baseline ecological and water resources management, environmental audit, land treatment of waste materials, and adjudicatory and expert witness consultation. He has served on numerous committees and councils and is an officer in a variety of professional organizations.

EDUARDO J. PADRÓN is president of Miami-Dade College (MDC), a learning-centered institution and the largest college in the nation, with a budget of over $500 million and 7,500 employees serving more than

168,000 students at six campuses. Dr. Padrón was named president of MDC in 1995. Prior to that, he was president of MDC s Wolfson campus from 1980 to 1995. Dr. Padrón received a PhD in economics from the University of Florida in 1970. He has drawn national respect for a broad range of innovations, beginning with successful programs for underserved and under-prepared students. Dr. Padrón has also played key leadership roles nationally through his service with the Carnegie Foundation, American Association of Community Colleges (AACC), American Council on Education (ACE), The College Board, American Association of Colleges and Universities (AACU), Higher Education Research and Development Institute (HERDI), League for Innovation in the Community College, and the national board of Campus Compact. He has been appointed to posts of national promi-nence by Presidents Carter, Bush, and Clinton and has received numerous honors throughout his career. His writings have appeared in many national journals, including his most recent on need- versus merit-based funding in *The College Board Review*. He is the recipient of the 2002 CEO of the Year Award from the Association of Community College Trustees (ACCT) and has received numerous other national and international awards, includ-ing those officially accorded by the governments of France, Spain, and Argentina. Locally and nationally, Eduardo Padrón continues to pursue his passion for opportunity and excellence in community-based education.

WILLIE PEARSON is professor of sociology at the Georgia Institute of Technology. Prior to joining the faculty at Georgia Tech in July 2001, he held a distinguished appointment as Wake Forest Professor of Sociology at Wake Forest University and Adjunct in Medical Education at Wake Forest University School of Medicine. Dr. Pearson received his PhD in sociology from Southern Illinois University at Carbondale in 1981. Dr. Pearson serves or has served on the editorial boards of *Sociological Spectrum*; *Science, Technology and Human Values*; *Journal of Science and Engineering Ethics*; and *Educational Evaluation and Policy Analysis*. Dr. Pearson serves or has served on committees, advisory boards and panels at the National Institutes of Health, National Science Foundation, American Chemical Society, Ameri-can Association for the Advancement of Science, Burroughs Wellcome Fund, Graduate Records Examination Board, Sloan Foundation, American Socio-logical Association, Sigma XI, and the National Research Council. He was elected president of the Mid-South Sociological Association(1987); a mem-ber of the Executive Council, American Sociological Association's Section on Science, Knowledge and Technology (1989-1991); and a governor of the National Conferences on Undergraduate Research (1994-2000). Dr. Pearson serves on the advisory committee for the National Academy of Engineering Center for the Advancement of Scholarship on Engineering Education and previously served as cochair of the National Research Council Committee for the Assessment of NIH Minority Research Training Programs.

SIDNEY A. RIBEAU is the sixteenth president of Howard University and the sixth African American to serve as its chief executive officer. Since taking office in August 2008, Dr. Ribeau has championed improving services to students through his Students First Campaign, strengthening research with emphasis in the STEM disciplines, enhancing the university's international footprint, and building upon a legacy of service. He was president of Bowling Green State University (BGSU) in Bowling Green, Ohio for 13 years before coming to Howard. Under his leadership, BGSU was recognized for its residential learning communities, values-based education, and innovative graduate programs. President Ribeau began his career in 1976 as a professor of communication studies at California State University, Los Angeles and later became chair of the University's Pan African Studies Department—a position he held until 1987, when he was named Dean of Undergraduate Studies at California State University, San Bernardino. He also held positions as dean of the College of Liberal Arts at California Polytechnic State University, San Luis Obispo, and vice president for Academic Affairs at California State Polytechnic University, Pomona. Dr. Ribeau serves on the boards of the Teachers Insurance and Annuity Association-College Retirement Equities Fund (TIAA-CREF), Worthington Industries, and the National Research Council Committee on Underrepresented Groups and the Expansion of the Science and Engineering Workforce Pipeline. He has served on the boards of the National Collegiate Athletic Association (NCAA), the United Way, the Regional Growth Partnership, the Andersons Inc. (Maumee, OH), and Convergys Corp. Some of his numerous honors include distinguished alumnus awards from Wayne State University and University of Illinois, scholarly recognition from the National Communication Association, and the President's Award from the National Association of Student Personnel Administrators. He received a BS degree from Wayne State University, and MA and PhD degrees in interpersonal and group communication from the University of Illinois, Urbana-Champaign.

JOHN BROOKS SLAUGHTER is president and CEO of the National Action Council for Minorities in Engineering (NACME), which provides leadership and support for the effort to increase the number of under-represented minorities in engineering. A former director of the National Science Foundation, chancellor of the University of Maryland, College Park, and president of Occidental College, Dr. Slaughter has a long and distinguished background as a leader in the education, engineering and the scientific communities. He is a member of the National Academy of Engineering (NAE), where he has served on the Committee on Minorities in

Engineering, chaired its Action Forum on Engineering Workforce Diversity, and is a current member of the NAE Council. Dr. Slaughter holds honorary degrees from more than 25 institutions, is the winner of the 1997 Martin Luther King Jr. National Award, and was also honored with the first U.S. Black Engineer of the Year award in 1987. He is the founding editor of the international journal *Computers & Electrical Engineering*.

RICHARD TAPIA is a mathematician and professor in the Department of Computational and Applied Mathematics at Rice University in Houston, Texas. He is internationally known for his research in the computational and mathematical sciences and is a national leader in education and outreach. Tapia's current Rice positions are University Professor, Maxfield Oshman Professor in Engineering, Associate Director of Graduate Studies, and Director of the Center for Excellence and Equity in Education. The first in his family to attend college, Tapia went on to receive BA, MA, and PhD degrees in mathematics from the University of California, Los Angeles. In 1967 he joined the Department of Mathematics at UCLA and then spent two years on the faculty at the University of Wisconsin. In 1970 he moved to Rice University, where he was promoted to associate professor in 1972 and full professor in 1976. He chaired the department from 1978 to 1983. He is currently an adjunct faculty member of both Baylor College of Medicine and the University of Houston. Tapia has authored or coauthored two books and more than a hundred mathematical research papers. Professor Tapia is recognized as a national leader in diversity and has delivered numerous invited addresses at national and international mathematics conferences, served on university diversity committees, and provided leadership at a national level. Richard Tapia's distinguished research accomplishments and service to the nation have brought him many honors. He was the first Hispanic elected to the National Academy of Engineering and one of the first appointed to the National Science Board, where he served from 1996 to 2002. He was also the first recipient of the Computing Research Association's A. Nico Habermann Award for outstanding contribution to aiding members of underrepresented groups within the computing research community. He was named one of the 20 most influential leaders in minority math education by the National Research Council; listed as one of the 100 most influential Hispanics in the United States by *Hispanic Business* magazine (2008); and given the Professor of the Year award by the Association of Hispanic School Administrators, Houston Independent School District, Houston, TX. In 2005, Tapia was elected to the Board of Directors for The Academy of Medicine, Engineering, and Science of Texas (TAMEST), which comprises Texas members of the National Academy of Engineering, National Academy of Sciences, and the Institute of Medicine. He received the National Science Foundation's inaugural Presidential Award for Excellence in Science, Mathematics, and Engineering Mentoring; the Lifetime Mentor Award from the American Association for the Advancement of Science; the Distinguished Service to the Profession Award from the Society for Industrial and Applied Mathematics; the Distinguished Public Service Award from the American Mathematical Society; the Distinguished Scientist Award from the Society for the Advancement of Chicanos and Native Americans in Science; and honorary doctorates from Carnegie Mellon University, Colorado School of Mines, and Claremont Graduate University. Two professional conferences have been named in his honor, recognizing his contributions to diversity: the Richard Tapia Celebration of Diversity in Computing Conference and the Blackwell-Tapia Conference, whose founders described Tapia as a seminal figure who inspired a generation of African American, Native American, and Latino/Latina students to pursue careers in mathematics.

LYDIA VILLA-KOMAROFF is the chief Executive officer at Cytonome. During her 20-year research career, Dr. Villa-Komaroff has held positions at the Massachusetts Institute of Technology, Harvard University, University of Massachusetts Medical School and Harvard Medical School. As a science administrator, she has been vice president for research at Northwestern University in Illinois and the vice president for research and chief operating officer of Whitehead Institute for Biomedical Research in Cambridge, MA. She also served as chair of the board of directors of Transkaryotic Therapies. In the United States, Dr. Villa-Komaroff's achievements have received national recognition. Profiled in the PBS special *DNA Detective*, Dr. Villa-Komaroff has been honored by the White House and is the recipient of three honorary doctorates. She is a member of the Hispanic Engineer National Achievement Hall of Fame and a fellow of the Association for Women in Science. She was named one of the 50 most important Hispanics in business and technology by *Hispanic Engineer and Information Technology* (2002) and one of the 100 most influential Hispanics by *Hispanic Business* Magazine (1997, 2003). As one of the country's most prominent Hispanic-American scientists, Dr. Villa-Komaroff is deeply committed to the recruitment and retention of women and minorities in science. She is a founding member of the Society for the Advancement of Chicanos and Native Americans in Science and has been both a board member and vice president of the organization. Dr. Villa-Komaroff received her PhD in cell biology from the Massachusetts Institute of Technology in 1975.

LINDA SUE WARNER is the president of Haskell Indian Nations University in Lawrence, Kansas. Warner is an accomplished educator. She was named winner of the 2001 Indian Educator of the Year Award by the National Indian Education Association for her lifelong dedication. Just last year, she was honored by the Pennsylvania State University's College of Education as winner of the Alumni Society's Leadership and Service Award. Warner has devoted 30 years to American Indian education policy and leadership. She has spread her pedagogy to numerous locations throughout the United States, beginning her career in the public schools of Missouri. In 1974, she entered the Bureau of Indian Affairs (BIA) in Alaska to serve as a teacher. She has been a faculty member at the University of Kansas, Pennsylvania State University, and the University of Missouri. She also served as a program director at the National Science Foundation. Most notable are her appointments to the National Advisory Council on American Indian Education (White House appointed) and the Department of Interior's Foundation for Excellence in American Indian Education. Her most recent position was with the Tennessee Board of Regents, the fifth largest university system in the country, where she served as associate vice chancellor for academic affairs.

Appendix D

Agendas for Public Meetings

MEETING ONE

Committee on Underrepresented Groups and the Expansion of
the Science and Engineering Workforce Pipeline
Keck Center of the National Academies
500 Fifth Street, NW,
Washington, DC 20001
Room 110
March 10-11, 2008

Monday, March 10, 2008

Continental breakfast available in the meeting room

Closed Session

8:00   Welcome and Overview of Agenda
        Freeman Hrabowski, Committee Chair

8:15   Charge to the Committee
        Freeman Hrabowski, Committee Chair

8:45   National Academies Discussion of Committee Balance and
             Potential Sources of Bias or Conflict of Interest
        Richard Bissell, Executive Director, Policy and Global Affairs,
        National Academies

Open Session

9:30   Welcome to Open Session
        Freeman Hrabowski, Committee Chair

9:35   Congressional Perspective on Issues and Study Charge
        Peter Ambler, Office of Congressman Silvestre Reyes
        Joye Purser, Office of Congresswoman Eddie Bernice Johnson

10:00   Break

10:15   Pathways to STEM Careers
        Willie Pearson, Georgia Institute of Technology, Workshop
            Report: *Broadening Participation Through a Comprehensive,
            Integrated System* (2005)

10:45   **CAWMSET and BEST**
        Beatriz Clewell, Urban Institute and former Executive Director,
            Congressional Commission on the Advancement of Women
            and Minorities in Science Engineering and Technology
            Development
        John Yochelson, President, Building Engineering and Science
            Talent

12:15   Break

Lunch available in meeting room

1:00   **Review of Data Trends**
        Lisa Frehill, Executive Director, Commission on Professionals in
            Science and Technology
        Earnestine Psalmonds, Senior Program Officer, National Research
            Council

2:15   **Federal Intervention Programs**
        Clifton Poodry, Division of Minority Opportunities in Research,
            NIGMS, National Institutes of Health
        Carl Person, Minority University Research and Education
            Programs, Office of Education, National Aeronautics and
            Space Administration
        Bernadette Hence, Minority Science and Engineering
            Improvement Program, Office of Postsecondary Education,
            U.S. Department of Education

3:45   Break

4:00   **Private Foundation Programs**
        Gail Christopher, Vice President for Programs, Health,
            W.K. Kellogg Foundation
        Peter Bruns, Vice President of Grants and Special Programs,
            Howard Hughes Medical Institute:
        Charles Terrell, Vice President, Division of Diversity Policy and
            Programs, American Association of Medical Colleges
        Mary Williams, Director of Communications and Administration,
            Gates Millennium Scholars Program, United Negro College
            Fund

5:15   Recess

Tuesday, March 11, 2008

Continental Breakfast will be available in the meeting room

Open Session

8:00   **Program Evaluation (NRC Reports for NIH)**
        Peter Henderson, National Research Council: NRC, *Evaluation
            of NIH Minority Research Trainings Programs*
        Adam Fagen, National Research Council: NRC, *Understanding
            Interventions That Encourage Minorities to Pursue Research
            Careers*

9:00   **Standing Our Ground I and II**
        Daryl Chubin, Director, AAAS Center for Advancing Science &
            Engineering Capacity

10:00   Break

10:15   **Demand for and Supply of Scientists and Engineers**
        Michael Teitelbaum, Vice President, Alfred P. Sloan Foundation

11:00   **Discussion of Schedule, Convocation, and Next Steps**

12:00   **Adjourn**

MEETING TWO

Joint Meeting of
Committee on Underrepresented Groups and the Expansion of
the S&E Workforce Pipeline
and
Committee on Capitalizing on the Diversity of the
S&E Workforce in Industry

National Academy of Sciences
2101 Constitution Avenue, NW
Washington, DC
Board Room
June 11-12, 2008

Wednesday, June 11, 2008
Joint Meeting of Committees

8:00   Breakfast available in the meeting room

8:30   **Welcome from the National Academies**
        Charles Vest, President, National Academy of Engineering
        Freeman Hrabowski, Chair, Committee on Underrepresented
            Groups and the Expansion of the S&E Workforce Pipeline
        Nicholas Donofrio, Co-Chair, Committee on Capitalizing on the
            Diversity of the S&E Workforce in Industry

9:00   **Perspectives from the U.S. Congress**
        U.S. Representative Silvestre Reyes

9:30   **Review of Study Committee Charges**
        Richard Bissell, Executive Director, Policy and Global Affairs,
            National Academies

10:00   Break

10:15   **Perspectives of Constituent Groups**
        Shirley Malcom, Head, Directorate for Education and Human
            Resources Programs, American Association for the
            Advancement of Science
        Edward M. Swallow, Chairman, National Security Science and
            Technology Workforce Division, National Defense Industrial
            Association

Jennifer McNelly, Vice President, National Center for the
American Workforce, National Association of Manufacturers

12:00 **Lunch Speaker**
Irma Arispe, Assistant Director for Life Sciences, White House
Office of Science and Technology Policy

1:00 **Perspectives from Federal Agencies**
Raynard Kington, Deputy Director, National Institutes of Health
Wanda Ward, Deputy Assistant Director, Education and Human
Resources Directorate, National Science Foundation
William Valdez, Director of the Office of Workforce Development
for Teachers and Scientists, U.S. Department of Energy

2:15 **S&E Labor Markets and Diversity: An Economic Perspective**
Sharon Levin, Research Professor of Economics, University of
Missouri-St. Louis
Mark Regets, Senior Analyst, Division of Science Resources
Statistics, National Science Foundation

3:15 Break

3:30 **Joint Planning Session**
Topics
Speakers

4:30 **Closing Remarks**
Freeman Hrabowski, Committee Chair
Nicholas Donofrio, Committee Chair

5:00 **Adjourn**

Thursday, June 12, 2008,
Committee on Underrepresented Groups and
the Expansion of the S&E Workforce Pipeline

Open Session

8:00   Breakfast available in the meeting room

8:30   **Recent Developments on Diversity in STEM Fields**
        "The Enhancing Science, Technology, Engineering, and
            Mathematics Education (E-STEM) Act of 2008"
        Edward Potosnak III, Einstein Fellow, Office of Congressman
            Michael M. Honda

Closed Session

9:30   **Committee Business**
        • Discussion of committee charge
        • Report outline
        • Information gathering: convocation and committee meetings
        • Information gathering: data, literature, reports, interviews,
          focus groups
        • Staff work plan
        • Next steps

10:30   Break

10:45   **Committee Business (continued)**

12:00   Working Lunch

1:00   **Adjourn**

MEETING THREE

Committee on Underrepresented Groups and
the Expansion of the Science and Engineering Workforce Pipeline
Keck Center of the National Academies
500 Fifth Street, NW
Washington, DC 20001
Room 206
October 22-23, 2008

Wednesday, October 22, 2008,
Room 206

Closed Session

1:00 **Welcome**
Freeman Hrabowski, Committee Chair

1:15 **Committee Discussion of Balance and Conflict of Interest**
Richard Bissell, Executive Director, Policy and Global Affairs,
The National Academies

Open Session

1:45 **Strategies for Minority Success in STEM**
Richard A. Tapia, University Professor, Maxfield-Oshman
Professor in Engineering, Director of the Center for
Excellence and Equity in Education, and Associate Director
of Graduate Studies, Rice University
Isiah Warner, Vice Chancellor, Office of Strategic Initiatives,
and Professor of Chemistry, Louisiana State University and
Agricultural and Mechanical College

2:45 Break

3:00 **Strategies for Minority Success in STEM (continued)**
Mary Ann Rankin, Dean of Natural Sciences and UTeach,
University of Texas at Austin
John Slaughter, President and CEO, National Action Council for
Minorities in Engineering, Inc.

4:00   **Power of Diversity**
Scott E. Page, Professor of Complex Systems, Political Science,
and Economics, University of Michigan and author of *The
Difference: How the Power of Diversity Creates Better
Groups, Firms, Schools, and Societies*

Closed Session

5:00   **Reception**
Keck Center, Room 1024

6:00   **Committee Dinner**
Keck Center, Room 1024

**Thursday, October 23, 2008,
Room 206**

Open Session

Continental Breakfast Available in the Meeting Room

8:30   **Welcome**
Freeman Hrabowski, Committee Chair

8:45   **Increasing Minority Participation and Success in K-12 STEM
Education**
Robert P. Moses, President, The Algebra Project, Inc.
Philip Sakimoto, Outreach and Diversity Specialist, Department
of Physics, University of Notre Dame
Joseph Merlino, Director, Math & Science Partnership of Greater
Philadelphia, Lasalle University

9:50   Break

10:00   **Minority Success in Undergraduate and Graduate STEM
Programs**

Gilda A. Barabino, Vice Provost for Academic Diversity, and
Professor and Associate Chair for Graduate Studies, Georgia
Institute of Technology

Aaron Velasco, President, Society for the Advancement of
Chicanos and Native Americans in Science (SACNAS), and
Associate Professor, Department of Geological Sciences,
University of Texas at El Paso

Valerie Petit Wilson, Associate Dean, Recruitment & Professional
Development, Brown University Graduate School; Clinical
Professor, Community Health, Brown University; and
Executive Director, Leadership Alliance

11:30 Data Issues
Mary Frase, Deputy Director, Division of Science Resources
Statistics, Directorate of Social, Behavioral, and Economic
Sciences, National Science Foundation

12:00 Committee Discussion

[Lunch will be available in the meeting room during this session]

1:00 Role of Minority Serving Institutions in Minority Success in
STEM Education
Norman C. Francis, President, Xavier University of Louisiana
Carol Davis, Tribal College Liaison, North Dakota EPSCoR, and
former Vice President, Turtle Mountain Community College

2:15 Break

Closed Session

2:30 Committee Discussion
Report from staff on data analyses and review of literature
Outline of Report: structure and recommendations
Agenda for meeting four

4:30 Adjourn

Appendix E

Recommendations on STEM Education from *Rising Above the Gathering Storm*

10,000 TEACHERS, 10 MILLION MINDS,
AND K–12 SCIENCE AND MATHEMATICS EDUCATION

Recommendation A:
*Increase America's talent pool by vastly improving
K–12 science and mathematics education.*

Implementation Actions

The highest priority should be assigned to the following actions and programs. All should be subjected to continuing evaluation and refinement as they are implemented.

Action A-1: Annually recruit 10,000 science and mathematics teachers by awarding 4-year scholarships and thereby educating 10 million minds.

Attract 10,000 of America's brightest students to the teaching profession every year, each of whom can have an impact on 1,000 students over the course of their careers. The program would award competitive 4-year scholarships for students to obtain bachelor's degrees in the physical or life sciences, engineering, or mathematics with concurrent certification as K–12 science and mathematics teachers. The merit-based scholarships would provide up to $20,000 a year for 4 years for qualified educational expenses, including tuition and fees, and require a commitment to 5 years of service in public K–12 schools. A $10,000 annual bonus would go to participating teachers in underserved schools in inner cities and rural areas. To provide the highest-quality education for undergraduates who want to become teachers, it would be important to award matching grants, on a one-to-one basis, of $1 million a year for up to 5 years, to as many as 100 universities and colleges to encourage them to establish integrated 4-year undergraduate programs leading to bachelor's degrees in the physical and life sciences, mathematics, computer sciences, or engineering *with teacher certification.* The models for this action are the UTeach and California Teach program.

Action A-2: Strengthen the skills of 250,000 teachers through training and education programs at summer institutes, in master's programs, and in Advanced Placement (AP) and International Baccalaureate (IB) training programs.

Use proven models to strengthen the skills (and compensation, which is based on education and skill level) of 250,000 *current* K–12 teachers.

•   *Summer institutes:* Provide matching grants to state and regional 1- to 2-week summer institutes to upgrade the skills and state-of-the-art knowledge of as many as 50,000 practicing teachers each summer. The material covered would allow teachers to keep current with recent developments in science, mathematics, and technology and allow for the exchange of best teaching practices. The Merck Institute for Science Education is one model for this action.

•   *Science and mathematics master's programs:* Provide grants to research universities to offer, over 5 years, 50,000 current middle school and high school science, mathematics, and technology teachers (with or without undergraduate science, mathematics, or engineering degrees) 2-year, part-time master's degree programs that focus on rigorous science and mathematics content and pedagogy. The model for this action is the University of Pennsylvania Science Teacher Institute.

•   *AP, IB, and pre-AP or pre-IB training:* Train an additional 70,000 AP or IB and 80,000 pre-AP or pre-IB instructors to teach advanced courses in science and mathematics. Assuming satisfactory performance, teachers may receive incentive payments of $1,800 per year, as well as $100 for each student who passes an AP or IB exam in mathematics or science. There are two models for this program: the Advanced Placement Incentive Program and Laying the Foundation, a pre-AP program.

•   *K–12 curriculum materials modeled on a world-class standard:* Foster high-quality teaching with world-class curricula, standards, and assessments of student learning. Convene a national panel to collect, evaluate, and develop rigorous K–12 materials that would be available free of charge as a *voluntary* national curriculum. The model for this action is the Project Lead the Way pre-engineering courseware.

Action A-3: Enlarge the pipeline of students who are prepared to enter college and graduate with a degree in science, engineering, or mathematics by increasing the number of students who pass AP and IB science and mathematics courses.

Create opportunities and incentives for middle school and high school students to pursue advanced work in science and mathematics. By 2010, increase the number of students who take at least one AP or IB mathematics or science exam to 1.5 million and set a goal of tripling the number who pass those tests to 700,000. Student incentives for success would include 50 percent examination fee rebates and $100 mini-scholarships for each passing score on an AP or IB science or mathematics examination. Although it is not included among the implementation actions, the committee also finds attractive the expansion of two approaches to improving K–12 science and mathematics education that are already in use:

- *Statewide specialty high schools:* Specialty secondary education can foster leaders in science, technology, and mathematics. Specialty schools immerse students in high-quality science, technology, and mathematics education; serve as a mechanism to test teaching materials; provide a training ground for K–12 teachers; and provide the resources and staff for summer programs that introduce students to science and mathematics.

- *Inquiry-based learning:* Summer internships and research opportunities provide especially valuable laboratory experience for both middle school and high school students.

BEST AND BRIGHTEST
IN SCIENCE AND ENGINEERING HIGHER EDUCATION

Recommendation C:
Make the United States the most attractive setting in which to study and perform research so that we can develop, recruit, and retain the best and brightest students, scientists, and engineers from within the United States and throughout the world.

Implementation Actions

Action C-1: Increase the number and proportion of U.S. citizens who earn bachelor's degrees in the physical sciences, the life sciences, engineering, and mathematics by providing 25,000 new 4-year competitive undergraduate scholarships each year to U.S. citizens attending U.S. institutions.

The Undergraduate Scholar Awards in Science, Technology, Engineering, and Mathematics (USA-STEM) would be distributed to states on the basis of the size of their congressional delegations and awarded on the basis of national examinations. An award would provide up to $20,000 annually for tuition and fees.

Action C-2: Increase the number of U.S. citizens pursuing graduate study in "areas of national need" by funding 5,000 new graduate fellowships each year.

NSF should administer the program and draw on the advice of other federal research agencies to define national needs. The focus on national needs is important both to ensure an adequate supply of doctoral scientists and engineers and to ensure that there are appropriate employment opportunities for students once they receive their degrees. Portable fellowships would provide a stipend of $30,000 annually directly to students, who would choose where to pursue graduate studies instead of being required to follow faculty research grants, and up to $20,000 annually for tuition and fees.

Action C-3: Provide a federal tax credit to encourage employers to make continuing education available (either internally or through colleges and universities) to practicing scientists and engineers.

These incentives would promote career-long learning to keep the workforce productive in an environment of rapidly evolving scientific and engineering discoveries and technological advances and would allow for retraining to meet new demands of the job market.

Appendix F

Ingredients for Success in STEM

There is no single pathway or pipeline in STEM education. Students start from diverse places, with different family backgrounds and schools and communities with different resources and traditions. There is substantial variation in mathematics and science education—particularly at the K-12 level across schools, districts, and states—with the range of variation reflecting everything from different approaches to teaching and learning mathematics in elementary school to the chasm between those who favor evolution and those who espouse creationism or intelligent design. STEM courses, moreover, may serve varied purposes for students on different tracks:

- Students differ in their fields of study—social sciences, psychology, mathematics, computer science, natural sciences, engineering—each of which has its own traditions, culture, educational progressions, and career paths.
- Students differ in terms of intended occupation, both by sector (academe, industry, nonprofit, government) and level of education (associate, bachelor's, master's, doctorate).
- Some students take mathematics and science courses never intending to major in a STEM field or work in a STEM occupation but nevertheless seek to be math and science literate.

All of this is to say, paths that start from places as varied as inner-city neighborhoods and wealthy suburbs and lead to jobs as divergent as those of a java programmer with a bachelor's degree and an academic biomedical researcher with a doctorate are very different paths even though we collectively group them as within the term STEM "pathway."

To assess the journey of underrepresented minorities in STEM education, a review of what it takes to become a scientist or engineer can set up a framework for understanding how to help underrepresented minorities navigate whatever STEM pathway they are on. While a set of pathways may be difficult to describe in detail, there are nonetheless ingredients for success in STEM that can be discussed, principally:

- The acquisition of knowledge, skills, and habits of mind;
- Opportunities to put these into practice;
- A developing sense of competence, confidence, and progress;
- Motivation to be in, a sense of belonging to, or self-identification with the field; and
- Information about stages, requirements, and opportunities.

These ingredients are present and require attention in some measure at every stage along the STEM educational continuum. Later, as one gets closer to entering the workforce, an additional ingredient may be a sense that, in a practical manner, the demands and benefits of the profession fit with one's lifestyle (e.g., provide a desired income level or work-life balance).

KNOWLEDGE, SKILLS, AND HABITS OF MIND

Knowledge, skills, and habits of mind are developed over time. Children enter elementary school as capable and generally enthusiastic science learners. *Taking Science to School* shows that children bring capabilities and prior knowledge that are "a resource that can and should be accessed and built upon during science instruction."[1] Cognitive researchers have determined that even young children in kindergarten possess strong reasoning skills. In combination with the knowledge already gained from their experiences and interactions with the natural world around them, this reasoning ability can be funneled into constructive science learning when in school. This science learning, then, should develop over the course of years in elementary and secondary school and postsecondary education as a "learning progression." Such a progression can be based on vertically articulated curricula in which units in higher grades build on units and concepts learned in the lower grades.[2] "Meaningful science learning takes time and learners need

---

[1] National Research Council. 2007.*Taking Science to School: Learning and Teaching Science in Grades K-8*. Washington, DC: The National Academies Press.

[2] J. Bruner. 1977. *The Process of Education*. Cambridge, MA: Harvard University Press, MA.

repeated, varied opportunities to encounter and grapple with ideas," *Ready, Set, Science!* asserts.[3]

What actually constitutes the content of the knowledge, skills, and habits of mind that students must acquire can be and is debated, but the general parameters can be briefly outlined. (As examples of science education by stage, Box F-1 presents four key "strands" in K-8 science education, illustrating one description of what students must acquire at that level, and Box F-2 presents a set of recommendations for undergraduate education in biology.)

In brief, the general parameters for STEM **knowledge** by broad field include:

- *Mathematics*: Basic facts and algorithms; algebra, trigonometry, geometry; problem solving ability; and verbal skills.[4]
- *Engineering*: Mathematical concepts, computational methods, science concepts, engineering design.[5]
- *Natural Sciences*: Facts, concepts, principles, laws, theories, and models of science. Facts cover specific areas of natural science (e.g., time, light waves, nature of force/velocity/acceleration, and theory of evolution)[6]
- *Social and Behavioral Sciences*: Sense of history and place; fundamentals of government and politics, economics, society, and human behavior.

Habits of mind, again by broad field, include:

- *Mathematics:* Thinking conceptually, logical reasoning, experimental thinking, inquisitiveness and the willingness to investigate, and the ability to take risks and accept failure.[7]
- *Engineering*: Systems thinking, creativity, optimism, collaboration, communication, and attention to ethical consideration.[8]
- *Natural Sciences*: Understanding of how concepts fit together, ability to generate and interpret evidence to build and refine models and explanations, use of mathematical reasoning, and employment of critical reasoning skills.[9]

---

[3] National Research Council. 2007. *Ready, Set, Science! Putting Research to Work in K-8 Science Classrooms.* Washington, DC: The National Academies Press.

[4] The College Board. College Board Standards for College Success: Mathematics and Statistics. 2006.

[5] National Academy of Engineering. 2009. *Engineering in K-12 Education.* Washington, DC: The National Academies Press.

[6] The College Board. *Science: College Board Standards for College Success* 2009. New York, NY: The College Board.

[7] *College Board Standards for College Success: Mathematics and Statistics.* ibid.

[8] National Academy of Engineering, *Engineering in K-12 Education,* ibid.

[9] National Research Council. 2007. *Ready, Set, Science!* Washington, DC: The National Academies Press and Science: College Board Standards, ibid.

BOX F-1
Four Key Strands in K-8 Science Education

Ready, Set, Science! describes four key strands to science education at the elementary level:

- *Understanding Scientific Explanations*: This strand involves learning the facts, concepts, principles, laws, theories, and models of science. However, it does so in a way that focuses on concepts and the links between them, rather than discrete facts. "To be proficient in science," the report argues, "students need to know, use, and interpret scientific explanations of the natural world. They must understand interrelations among central scientific concepts and use them to build and critique scientific arguments."
- *Generating Scientific Evidence*: This strand entails generating and evaluating evidence to build and refine models and explanations, design and analyze investigations, and construct and defend arguments. This strand also involves mastering the conceptual, mathematical, physical, and computational tools that are needed to construct and evaluate claims.
- *Reflecting on Science Knowledge*: Proficient science learners understand that scientific knowledge builds on itself and can be revised over time. Students recognize that predictions or explanations can be revised on the basis of seeing new evidence, learning new facts, or developing a new model.
- *Participating Productively in Science*: Science is a social enterprise. Proficiency entails participation in a scientific community—at this level, the classroom—and mastery of productive ways to present scientific information and arguments and work with their peers in carrying out investigations.

In this model, science learning can be based on the way real scientists do science, and content and process interact as students move toward proficiency.

SOURCE: National Research Council. 2007. *Ready, Set, Science! Putting Research to Work in K-8 Science Classrooms.* Washington, DC: The National Academies Press.

- *Social Sciences*: Theoretical understanding, ability to organize information to test and refine a theory.

With some varying degree by field, the additional **skills** needed for STEM success include:

- Persistence;
- Reading, writing, and communication;
- Basic mathematical skills, including the ability to do word problems;
- Ability to analyze and interpret statistical data;

BOX F-2
Vision and Change in Undergraduate Biology Education: A Call to Action

Most faculty agree that to be scientifically literate, students need to understand a few overarching core concepts: evolution; pathways and transformations of energy and matter; information flow, exchange, and storage; structure and function; and systems. As important, undergraduates need to understand the process of science, the interdisciplinary nature of the new biology, and how science is closely integrated within society. Students also should be competent in communication and collaboration, as well as have a certain level of quantitative competency, and a basic ability to understand and interpret data. These concepts and competencies should be woven into the curriculum and reinforced throughout all undergraduate biology coursework.

Student-Centered Classrooms and Learning Outcomes: In practice, student-centered classrooms tend to be interactive, inquiry-driven, cooperative, collaborative, and relevant. Classes authentically mirror the scientific process, convey the wonder of the natural world and the passion and curiosity of scientists, and encourage thinking. In addition, classes include both formal and informal assessment and regular feedback to students and faculty to help inform teaching and monitor student learning. And finally, regardless of their majors and eventual careers, students should have opportunities to participate in authentic research experiences and learn how to evaluate complex biological problems from a variety of perspectives, not just recite facts and terminology.

Understanding Key Concepts and Competencies: To be current in biology, students should also have experience with modeling, simulation, and computational and systems-level approaches to biological discovery and analysis, as well as with using large databases. Having a basic understanding of core concepts that form the very basis of life on earth, combined with training in newer approaches to biological research, provides students with insights into the process of scientific discovery, as they develop the tools they will need to succeed in tomorrow's classrooms and board rooms.

Strategies for Change: To ensure a smooth transition to student-centered teaching and learning in undergraduate biology courses, all biology faculty and tenure review committees need to insist that the academic reward system value teaching and mentoring, set clear and concrete guidelines for assessment of these activities, and incorporate regular, formative and adaptive assessment of teaching effectiveness. Faculty need to come to consensus on the overarching, central concepts of biology that should be taught within their division or department, and define learning outcomes for those key concepts so that all faculty are working together toward the same learning goals as students move through their department.

The ultimate goal for biology departments should be to develop and grow communities of scholars at all levels of the educational process—from undergraduates to faculty to administrators—all committed to creating, using, assessing, and disseminating effective practices in teaching and learning. This kind of department-wide implementation requires cultural changes by all stakeholders and a commitment to elevate the scholarship of teaching and learning within the discipline as a professional activity.

SOURCE: National Science Foundation and American Association for the Advancement of Science, *Vision and Change in Undergraduate Education: A Call to Action*, a summary of recommendations made at a national conference, July 15-17, 2009.

- Ability to use scientific method; and
- Orientation toward learning, good study skills, and ability to take responsibility for one's own education.

Recently, national dialogue regarding "twenty-first century skills" suggests that, in addition to deep knowledge of the substance of a field, graduates at the bachelor's level and above also need professional skills that may include communication, project management, and ability to work in teams, proficiency in the use of computers, critical thinking, customer awareness, entrepreneurship, ethics, and regulation.[10]

PRACTICE, LEARNING, AND COMPETENCE

Opportunities to put knowledge, skills, and habits of mind into practice serve two important purposes. First, through inquiry-based learning or engineering design activities, students use and create scientific and technical knowledge, come to understand concepts, learn how to generate evidence that can be used to build and refine models and explanations, and develop an appreciation for reflection on experimental outcomes and the way they shape our knowledge. Second, through research or design activities, students also develop a sense of competence in mathematics, science, and technology. Competence is critical to identification with a field of endeavor such as STEM. There is significant attrition from STEM majors at the end of the freshman year in college, and research has shown, for example, that those who switch tended to blame themselves and their abilities when they encountered difficulties, while those who persisted tended instead to blame an external cause, such as the professor, a teaching assistant, or available laboratory resources.[11] A sense of competence is also significantly related to persistence, which, especially in mathematics, is critical to success.[12]

INTEREST, MOTIVATION, BELONGING, AND SELF-IDENTIFICATION

Beyond providing threshold education and higher-level preparation for STEM pathways, schools can also identify and encourage students who are **motivated** in mathematics and science to more fully develop their knowledge base and potential. Programs can include efforts to place students

---

[10] National Research Council. 2008. *Science Professionals: Master's Education for a Competitive World*. Washington, DC: The National Academies Press.

[11] Seymour, Elaine, Nancy M. Hewitt. 1997. *Talking About Leaving: Why Undergraduates Leave the Sciences*. Boulder, CO: Westview Press.

[12] Gladwell's Outliers: Timing is Almost Everything. Available at http://www.businessweek.com/magazine/content/08_48/b4110110545672.html.

in science and mathematics magnet schools or to encourage enrollment in Advanced Placement (AP), International Baccalaureate (IB), or similar advanced courses. Such participation in AP, for example, has correlated with higher rates of college enrollment and success.[13]

In college, students continue to grow along the STEM pathway. They continue to acquire knowledge both broadly and in their intended STEM field. It is important to continue to nurture **interest** in science and engineering as students continue on this pathway. Traditionally, many introductory courses in the sciences have functioned to "weed out" students rather than to encourage them. Research has shown, however, that these courses are more likely to weed out those who do not like the competitive culture of science than those who are not good at it. These students who switch majors could contribute in STEM if they were encouraged and nurtured in their interest instead.[14]

Engagement in rich research experiences allows for the further development of interest in, competence in, and identification with STEM. Research has shown that these experiences with the operations of science very often seize the interest of students who then develop a fascination that translates into a career in STEM. In addition, summer programs in mathematics, science, and engineering that include or target minority high school and undergraduate students provide experiences that stimulate interest in these fields through study, active research or projects, and the development of a cadre of students who support each other in their interests. Similarly, providing opportunities for students to engage in professional development activities, particularly in graduate programs, will provide additional opportunities to both develop the student and socialize them within a discipline and profession. These activities include opportunities for networking, participation in conferences, and presentations of research (on campus or in other professional settings).

Even if students are prepared, have adequate information, and are ambitious and talented enough to succeed in STEM fields, success may also hinge on the extent to which students feel socially and intellectually integrated into their academic programs and campus environments. The importance of social and intellectual integration for success is critical to all students, regardless of background. For minority students who may feel, or be made to feel, like outsiders as they see few others "like themselves" among the student and faculty populations, this issue takes on even greater salience. The development of peer-to-peer support, study groups, program activities fostering social integration, and tutoring and mentoring programs may go a long way to overcome this critical hurdle (Astin 1993, Kuh 2003,

---

[13] The 6th Annual AP Report to the Nation. 2010. The College Board, ibid.
[14] Seymour and Hewitt, ibid.

Tinto 1993, Pike and Kuh 2005, Swail 2003). Higher education programs should also develop "bridging programs" that assist students as they move across transition points. These programs include a focus on preparation for the next level, guidance from mentors on mastering the transition, the development of connections between programs, and financial support as necessary.

The issue of self-efficacy cannot be ignored. Bandura contends that self-efficacy beliefs impact every aspect of students' lives and can powerfully influence the level of accomplishment that they ultimately experience. Students form their self-efficacy perceptions by interpreting information from four sources: mastery experience, vicarious experience, social persuasions, and physiological reactions. For most, mastery experience is the most influential. Success raises self-efficacy; failure lowers it.

AWARENESS AND INFORMATION

In primary school—and continuing into middle and high school years—developing an **awareness** of STEM careers can provide inspiration for students that can be reinforced in mathematics and science courses. School districts can introduce students to STEM careers, starting even in preschool, through awareness activities that would include speakers (role models), activities, field trips, participation in science or engineering programs, and links to summer programs. Employers can form partnerships with K-12 schools to promote STEM education and careers to minority students. They can also provide STEM employees who can serve as role models or mentors and they can provide internships that connect for students the worlds of science and work.[15] Higher education institutions could engage in outreach and recruitment activities, in particular considering the development of targeted outreach programs that constitute a "feeder system" for their institutions. The federal government could engage in a marketing campaign designed to "change the face" of STEM careers in the public eye, and especially for families who play an important role in shaping the notions of what their children can become.[16]

Many students have insufficient **information** about educational and career opportunities and options, both in general and for STEM, at critical decision points in middle and high school. There may be few opportunities to learn about these options unless institutions—schools, churches, community groups—make an effort to provide role models and information. To complement efforts to raise awareness of STEM careers generally,

---

[15] Confronting the New American Dilemma. 2008. National Action Council for Minorities in Engineering, Inc.

[16] Diversity and Innovation Caucus Stakeholders' Listening Meeting, February 28, 2008.

counseling in middle and high schools can provide important and timely information in a practical way about what is academically necessary—in high school and in college—to pursue STEM careers. This counseling can also focus on preparing students and families for their initial interactions with higher education institutions, including the application and financial aid processes.

Very often, in underresourced schools—ones that are often predominantly minority— students are not encouraged to take the next level of courses needed for college preparation. A recent College Board report argues, "Curriculum rigor trumps just about everything else in predicting college success" and then goes on to note further that "No ethnic group in America comes close to attending high schools in which a college-prep curriculum is universally available. Minority students and those from low-income families have the least access to such a curriculum."[17]

In these cases, a program such as the Algebra Project, which encourages student interest in and demand for quality secondary instruction in mathematics and then provides multifaceted intervention, can help overcome this critical obstacle. Students who see achievement in mathematics as both a right and a door to opportunity have an increased probability of success.

Advising and mentoring are also important to provide support and information, both in general and at critical decision points. For undergraduates, academic advising about and support for preparation and application for graduate school can make the difference between whether a student continues in the STEM pathway. In graduate school, mentors provide important guidance and support to students, reducing attrition, helping students maximize their educational experience, and providing guidance on launching a career. Higher education institutions can develop faculty who will serve as strong, engaged mentors for STEM students generally and for minority STEM students in particular.

INSTITUTIONAL INGREDIENTS

Although it is important that each individual student have access to the ingredients for success described above, there is also a set of institutional preconditions that affect all of these requirements for success in STEM education. They include qualified teachers who have strong scientific knowledge and understand how students learn; strong mathematics, science, and engineering curricula that provide knowledge, skills, and habits of mind; an institutional setting designed to provide or support each of the

---

[17] College Board. 2008. *Coming to Our Senses: Education and the American Future.*

requirements and time to achieve them; counseling and mentoring, much of it stage-specific, that helps the student navigate the path; the financial and social support students need to sustain them; and the availability or accessibility of institutional research infrastructure—that is, laboratories and equipment.

Appendix G

Baccalaureate Origins of Underrepresented Minority PhDs

Appendix G examines in detail the national data on baccalaureate origins of African Americans and Hispanics who earn PhDs in the natural sciences and engineering. The focus is on baccalaureate origins of PhDs because the analysis examines the production of the broad array of baccalaureate institutions rather than just doctoral institutions, yet it is centered on the preparation of students who go on to earn doctorates.

This is not meant to imply that preparation of bachelor's- and master's-level scientists and engineers is not important. Perhaps it is more important now than ever, with innovations such as the professional science master's degree spreading nationwide. This is not meant either to let the elite research universities off the hook when it comes to accepting and graduating underrepresented minority doctoral students in science and engineering—they are definitely responsible, particularly since their production has been collectively inadequate in graduating underrepresented minorities with doctorates in STEM fields. This is meant, however, to suggest that holding students to high standards and expectations such as one would find in the preparation of students for success at the doctoral level is very important, and so it is instructive to learn more about who is doing that and in what ways.

CONTEXT: SCIENCE AND ENGINEERING

The National Science Foundation (NSF) reports in *The Role of HBCUs as Baccalaureate-Origin Institutions of Black S&E Doctorate Recipients* that African American S&E doctorate recipients *earned their bachelor's degrees from a wide range of institutions.*[1] In 2006, one-third of these new doctorates had earned their bachelor's from a Historically Black College or University (HBCU) and two-thirds from non-HBCUs. Similarly, about 30 percent of the undergraduate institutions awarding bachelor's degrees to these individuals were HBCUs. Another 25 percent were non-HBCU research universities, and the rest of the institutions were from a range of non-HBCUs, including doctorate, master's, and liberal arts colleges, as well as a group of foreign institutions.

The proportion of African American S&E doctorates who had received their bachelor's degrees from HBCUs has fluctuated in recent decades, as NSF relates:

> In the latter 1970s, over 40 percent of black S&E doctorate recipients received their baccalaureate degrees from HBCUs. This percentage fell to 25 percent in the first part of the 1990s before increasing to about 33 percent in 2006. During the same period (1977-2006), the share of blacks receiving bachelor's degrees from HBCUs fell from 36 percent to 21 percent."[2]

But the role of HBCUs is strong in terms of overall numbers per institution. While they award a minority of the bachelor's degrees to African American S&E doctorates, the institutions awarding the largest number of bachelor's to this group are HBCUs. NSF reports that for African American S&E doctorate recipients in the period 1997-2006, the top 8 baccalaureate-origin institutions were HBCUs, and overall, 20 of the top 50 baccalaureate institutions were. The top 5 baccalaureate institutions were Howard University, Spelman College, Hampton University, Florida A&M University, and Morehouse College.

When normalized for the number of bachelor's degrees awarded nine years earlier to African American undergraduates, however, another important picture emerges. In this case, only 5 of the top 50 baccalaureate institutions for 1997-2006 African American S&E doctorates, including the social sciences, are HBCUs, with just Spelman in the top 25. The top five institutions were Massachusetts Institute of Technology, Swarthmore College, Princeton University, Harvard University, and Amherst College. This shows the role that elite predominantly white institutions (PWIs) can play. The "normalization" masks that these institutions have actually produced just small numbers that are a relatively high percentage relative to a small base and primarily in the social sciences (with the exception of MIT), but they

---

[1] Joan Burrelli and Alan Rapoport, "Role of HBCUs as Baccalaureate-Origin Institutions of Black S&E Doctorate Recipients," InfoBrief (NSF 08-319), National Science Foundation, Science Resources Statistics, August 2008.

[2] Ibid.

indicate the potential that could be unleashed when these institutions attract greater numbers of underrepresented minorities in STEM.

ANALYSIS: FOCUS ON NATURAL SCIENCE AND ENGINEERING

Our further analysis focuses more specifically on the natural sciences and engineering (NS&E) for the period 2002-2006 (most recent five years studied) to explore institutional strengths in particular areas.

As shown in Table G-1, the top 10 baccalaureate institutions of African Americans who went on to earn doctorates in the natural sciences and engineering (NS&E) for the period 2002-2006 were HBCUs. This is a pattern similar to the NSF analysis for science and engineering, though with more HBCUs rising into the top 10 group. The top baccalaureate institutions for NS&E doctorates were Florida A&M University, Howard University, Hampton University, North Carolina A&T State University, Spelman College, Morehouse College, Southern University at Baton Rouge, Xavier University of Louisiana, Tuskegee University, and Morgan State University.

If we expand the analysis to the "top 25" institutions (actually 28, as there is a 4-way tie for the 25th spot), 15 of these institutions were HBCUs and 13 were non-HBCUs. The highest ranked PWIs were the University of Maryland Baltimore County, Massachusetts Institute of Technology, University of Maryland College Park, North Carolina State University, Georgia Institute of Technology, and the City University of New York (CUNY) City College.

Historically Black Institutions

The ranking of the 15 HBCUs in NS&E is higher than the ranking of the 11 HBCUs in the top 25 baccalaureate institutions for African Americans who earned doctorates in all S&E fields during this period. Florida A&M University and North Carolina A&T State University, in particular, ranked higher in the NS&E list than they did in the overall S&E list because of their strong engineering programs. As shown in Table G-2, burrowing down further into specific NS&E fields, distinctive patterns of institutional focus emerge:

- Three institutions have large numbers due to **engineering** programs that produce bachelor's who go on to doctorates in the field. These include one HBCU (North Carolina A&T State University) and two PWIs (Massachusette Institute of Technology and Georgia Institute of Technology).
- Two institutions have similar concentrations in the life sciences, particularly the **biological sciences**. These are Xavier University, an HBCU,

TABLE G-1 Top 25 Baccalaureate Origin Institutions of African American Doctorates in the Natural Sciences and Engineering (NS&E), 2002-2006

Rank	Institution	2002	2003	2004	2005	2006	Total
	Total	399	386	464	495	488	2232
1	Florida Agricultural and Mechanical University	3	6	10	19	13	51
2	Howard University	10	5	7	12	14	48
3	Hampton University	9	6	11	7	11	44
4	North Carolina Agricultural & Tech State Univ	6	6	12	8	10	42
4	Spelman College	13	5	7	9	8	42
6	Morehouse College	3	7	7	9	9	35
7	Southern University A&M Col at Baton Rouge	4	6	7	11	5	33
8	Xavier University of Louisiana	1	4	12	10	5	32
9	Tuskegee University	9	4	5	4	9	31
10	Morgan State University	3	7	7	4	9	30
11	University of Maryland Baltimore County	2	3	6	6	7	24
12	Massachusetts Institute of Technology	3	6	6	3	3	21
12	University of Maryland at College Park	1	5	6	5	4	21
14	Alabama Agricultural and Mechanical University	.	4	2	6	7	19
14	North Carolina State University at Raleigh	3	3	3	4	6	19
16	Georgia Institute of Technology, Main Campus	1	2	7	4	3	17
16	Jackson State University	3	3	2	8	1	17
18	CUNY City College	3	2	2	3	6	16
18	Tougaloo College	2	6	2	1	5	16
20	Norfolk State University	3	3	2	4	3	15
20	University of North Carolina at Chapel Hill	2	2	5	5	1	15
21	Prairie View A&M University	2	1	7	1	3	14
21	Princeton University	3	4	3	1	3	14
21	University of Virginia, Main Campus	4	2	1	3	4	14
25	Cornell University, All Campuses	2	2	1	4	4	13
25	University of Florida	1	4	3	3	2	13
25	University of Pennsylvania	2	3	3	1	4	13
25	University of South Carolina at Columbia	6	2	1	2	2	13

NOTES:
The years 2002-2006 are the five most recent years of available data.
The table includes 28 institutions, as 4 were tied for 25th.
The African Americans included in this table are U.S. citizens or permanent residents.

KEY:

Historically Black Colleges and Universities
Predominantly White Colleges and Universities

SOURCE: NSF/SRS, WebCASPAR (Survey of Earned Doctorates).

and the University of Maryland Baltimore County, PWI with a program that has focused on the development of minorities in the biological sciences. Hampton University, noted below, also has a large concentration in the biological sciences.

• Alabama A&M University, Jackson State University, and Southern University at Baton Rouge have concentrations more generally in the life sciences, with a focus on the agricultural sciences. (The four top schools in the agricultural sciences are these three, plus Tuskegee University, noted below.)

• Eight institutions have concentrations in both engineering and the life sciences. These include five HBCUs (Florida A&M University, Howard University, Morgan State University, Tuskegee University, and Prairie View A&M University) and three PWIs (University of Maryland College Park, North Carolina State, and City University of New York City College).

• Hampton, Spelman, Morehouse, Norfolk State, and Tougaloo have granted bachelor's degrees to future doctorates across the **natural sciences** disciplines.

Thus, for HBCUs, there are several strategies evident for developing African Americans who earn doctorates in NS&E fields: Five baccalaureate institutions educate undergraduates across a range of natural sciences disciplines; six institutions have strong engineering programs; one has a particularly strong program in the biological sciences; four have strong programs in the agricultural sciences; and a handful have strong programs in both engineering and the life sciences.

Predominantly White Institutions

Among the top six PWIs, there are three noteworthy approaches that can be discerned in Table G-3:

• MIT admits outstanding African American engineering students, a small number who have a higher propensity for graduating and continuing on to doctoral study than other African American undergraduates.

• Georgia Tech and UMBC have focused efforts to recruit, support, and graduate students in engineering and the biological sciences who continue on to doctoral study in their fields.

• North Carolina State University, University of Maryland College Park, and City University of New York City College have relatively large numbers of African American students who are distributed across fields, some small percentage of whom continue to graduate school.

Taken individually, the achievements of MIT, Georgia Tech, and UMBC are remarkable. Taken together, unfortunately, the numbers they are pro-

TABLE G-2 Top 15 Baccalaureate Origin Institutions of African American Doctorates in the Natural Sciences and Engineering (NS&E) that are Historically Black Colleges and Universities (HBCUs), by Broad Field, 2002-2006 (most recent 5 years)

HBCUs	Total
Florida Agricultural and Mechanical University	51
Howard University	48
Hampton University	44
North Carolina Agricultural & Tech State University	42
Spelman College	42
Morehouse College	35
Southern University A&M Col at Baton Rouge	33
Xavier University of Louisiana	32
Tuskegee University	31
Morgan State University	30
Alabama Agricultural and Mechanical University	19
Jackson State University	17
Tougaloo College	16
Norfolk State University	15
Prairie View A&M University	14
TOTAL HBCUs	469

TABLE G-2 Top 15 Baccalaureate Origin Institutions of African American Doctorates in the Natural Sciences and Engineering (NS&E) that are Historically Black Colleges and Universities (HBCUs), by Broad Field, 2002-2006 (most recent 5 years) (continued)

HBCUs	Total
Florida Agricultural and Mechanical University	51
Howard University	48
Hampton University	44
North Carolina Agricultural & Tech State University	42
Spelman College	42
Morehouse College	35
Southern University A&M Col at Baton Rouge	33
Xavier University of Louisiana	32
Tuskegee University	31
Morgan State University	30
Alabama Agricultural and Mechanical University	19
Jackson State University	17
Tougaloo College	16
Norfolk State University	15
Prairie View A&M University	14
TOTAL HBCUs	469

TABLE G-3 Top 13 Baccalaureate Origin Institutions of African American Doctorates in the Natural Sciences and Engineering (NS&E) that are Predominantly White Universities, by Broad Field, 2002-2006 (most recent 5 years)

Non-HBCUs	Total
University of Maryland Baltimore County	24
Massachusetts Institute of Technology	21
University of Maryland at College Park	21
North Carolina State University at Raleigh	19
Georgia Institute of Technology, Main Campus	17
CUNY City College	16
University of North Carolina at Chapel Hill	15
Princeton University	14
University of Virginia, Main Campus	14
Cornell University, All Campuses	13
University of Florida	13
University of Pennsylvania	13
University of South Carolina at Columbia	13
TOTAL Non-HBCUs	213

TABLE G-3 Top 13 Baccalaureate Origin Institutions of African American Doctorates in the Natural Sciences and Engineering (NS&E) that are Predominantly White Universities, by Broad Field, 2002-2006 (most recent 5 years) (continued)

Non-HBCUs	Total
University of Maryland Baltimore County	24
Massachusetts Institute of Technology	21
University of Maryland at College Park	21
North Carolina State University at Raleigh	19
Georgia Institute of Technology, Main Campus	17
CUNY City College	16
University of North Carolina at Chapel Hill	15
Princeton University	14
University of Virginia, Main Campus	14
Cornell University, All Campuses	13
University of Florida	13
University of Pennsylvania	13
University of South Carolina at Columbia	13
TOTAL Non-HBCUs	213

ducing do not move the overall trend-line in numbers of African American doctorates very much.

TOP BACCALAUREATE INSTITUTIONS OF HISPANIC NS&E DOCTORATES

Context: Science and Engineering

The NSF has not produced an *InfoBrief* on Hispanic-serving institutions similar to the one it has produced on Historically Black Colleges and Universities, so we are not able to provide the same data on the larger S&E context that we did for African Americans.

Analysis: Focus on Natural Sciences and Engineering

As shown in Table G-4, of the top 25 baccalaureate institutions (actually 26, as two are tied for 25th) of Hispanics who earned doctorates in the natural sciences and engineering (NS&E) during the period 2002-2006, not surprisingly, three are campuses of the University of Puerto Rico. The remaining institutions include 17 predominantly white institutions (PWIs) and 5 Hispanic-Serving Institutions (HSIs).

Among the top 25 baccalaureate institutions for African American NS&E doctorates, just 13 are PWIs, so the composition of the largest baccalaureate institutions, particularly for Hispanic NS&E students who are not from Puerto Rico, differs substantially from those that provide baccalaureate education for African Americans.

Also, in contrast to the position of HBCUs as baccalaureate institutions of African American NS&E doctorates, the largest baccalaureate institutions for Hispanics after the University of Puerto Rico campuses are PWIs. The largest institutions are: University of Puerto Rico Mayaguez, University of Puerto Rico, Rio Piedras, University of California Berkeley, Massachusetts Institute of Technology, and the University of Florida.

The baccalaureate institutions of Hispanic NS&E doctorates are geographically concentrated. Not surprisingly, the majority (14) of the top institutions are located in the West (California, Texas, New Mexico, and Arizona). This geographic concentration underscores an important difference between HBCUs, which were created with the purpose of educating African Americans, and HSIs, which, with a small number of exceptions, do so because they are located in or near large Hispanic populations.

The students who attend the top 25 institutions are predominantly in the life sciences, followed by engineering. Some patterns emerge from an examination of Table G-5 which provides data by broad field for the top institutions:

TABLE G-4 Top 25 Baccalaureate Origin Institutions of Hispanic Doctorates in the Natural Sciences and Engineering (NS&E), 2002-2006

Rank	Institution	2002	2003	2004	2005	2006	Total
	Total	376	377	384	434	467	2038
1	University of PR Mayaguez Campus	40	22	28	29	49	168
2	University of PR Rio Piedras Campus	25	20	20	27	22	114
3	University of California-Berkeley	11	13	9	7	11	51
4	Massachusetts Institute of Technology	8	12	9	2	5	36
4	University of Florida	2	8	13	5	8	36
6	University of California-Davis	4	9	6	10	6	35
6	University of PR Humacao University College	5	6	3	8	13	35
8	University of Texas at Austin	2	8	9	5	10	34
8	University of Texas at El Paso	5	7	4	6	12	34
10	University of California Los Angeles	5	6	8	4	10	33
11	Texas A&M University Main Campus	8	5	5	3	9	30
12	New Mexico State University, All Campuses	2	5	5	7	9	28
13	University of California-San Diego	6	5	4	7	5	27
14	Florida International University	3	7	3	9	4	26
14	University of Miami	5	3	7	5	6	26
16	University of California-Irvine	1	5	7	7	5	25
17	University of New Mexico, All Campuses	3	3	6	6	6	24
18	Cornell University, All Campuses	5	4	4	4	5	22
18	University of Illinois at Urbana-Champaign	7	5	4	4	2	22
20	University of Arizona	2	1	7	6	5	21
20	University of Texas at San Antonio	2	5	2	6	6	21
22	University of California-Santa Cruz	5	4	4	4	3	20
23	Rutgers the State Univ of NJ New Brunswick	2	2	5	5	5	19
24	Stanford University	3	7	4	3	1	18
25	Harvard University	1	1	5	8	2	17
25	Princeton University	4	3	8	1	1	17

NOTES:

The years 2002-2006 are the five most recent years of available data.

The table includes 26 institutions as 2 were tied for 25th.

The Hispanics included in this table are U.S. citizens or permanent residents.

KEY:

Puerto Rico
California
Texas
New Mexico or Arizona
Florida

SOURCE: NSF/SRS, WebCASPAR (Survey of Earned Doctorates).

TABLE G-5 Top 25 Baccalaureate Origin Institutions of Hispanic Doctorates in the Natural Sciences and Engineering (NS&E), by Broad Field, 2002-2006 (most recent 5 years)

Institutions	Total
University of Puerto Rico Campuses	
University of PR Mayaguez Campus	168
University of PR Rio Piedras Campus	114
University of PR Humacao University College	35
Hispanic-Serving Institutions	
University of Texas at El Paso	34
New Mexico State University, All Campuses	28
Florida International University	26
University of Miami	26
University of New Mexico, All Campuses	24
University of Texas at San Antonio	21
Other Institutions	
University of California-Berkeley	51
Massachusetts Institute of Technology	36
University of Florida	36
University of California-Davis	35
University of Texas at Austin	34
University of California-Los Angeles	33
Texas A&M University Main Campus	30
University of California-San Diego	27
University of California-Irvine	25
Cornell University, All Campuses	22
University of Illinois at Urbana-Champaign	22
University of Arizona	21
University of California-Santa Cruz	20
Rutgers the State Univ of NJ New Brunswick	19
Stanford University	18
Harvard University	17
Princeton University	17
Total	939

TABLE G.5 Top 25 Baccalaureate Origin Institutions of Hispanic Doctorates in the Natural Sciences and Engineering (NS&E), by Broad Field, 2002-2006 (most recent 5 years) (continued)

Institution	Total
University of Puerto Rico Campuses	
University of PR Mayaguez Campus	168
University of PR Rio Piedras Campus	114
University of PR Humacao University College	35
Hispanic Serving Institutions	
University of Texas at El Paso	34
New Mexico State University, All Campuses	28
Florida International University	26
University of Miami	26
University of New Mexico, All Campuses	24
University of Texas at San Antonio	21
Other Institutions	
University of California-Berkeley	51
Massachusetts Institute of Technology	36
University of Florida	36
University of California-Davis	35
University of Texas at Austin	34
University of California-Los Angeles	33
Texas A&M University Main Campus	30
University of California-San Diego	27
University of California-Irvine	25
Cornell University, All Campuses	22
University of Illinois at Urbana-Champaign	22
University of Arizona	21
University of California-Santa Cruz	20
Rutgers the State Univ of NJ New Brunswick	19
Stanford University	18
Harvard University	17
Princeton University	17
Total	939

For Puerto Rico, the Mayaguez campus has a strong **engineering** program. The Mayaguez and Rio Piedras campuses both have strong **life sciences** programs.

For the continental United States, three institutions have strong **engineering** programs. One is an HSI (University of Texas at El Paso) and two are PWIs (MIT and Texas A&M). MIT is the one institution that appears on the list of top baccalaureate institutions for both African American and Hispanic PhDs, both rankings due to engineering programs, though MIT also has a strong record as the baccalaureate institution for Hispanics in the **physical sciences**.

In the **life sciences**, the largest institutions are two PWIs—University of California-Berkeley and the University of Florida—that have a particular focus in this broad field. The next three largest programs are at New Mexico State (an HSI) and the University of California-Davis and the University of Texas at Austin (both PWIs). For New Mexico State, about half of the students in the life sciences are in the biological sciences, and the other half are in the agricultural sciences. The other programs focus more strongly on the biological sciences.

The rest of the institutions not discussed already awarding bachelor's are also largely concentrated in the **life sciences**. These include the University of California-Los Angeles, Florida International University, University of California-San Diego, University of Miami, Cornell, University of California-Irvine, University of California- Santa Cruz, Rutgers, Stanford, Harvard, Princeton, and the Universities of New Mexico, Arizona, Illinois, and Texas at San Antonio.

Appendix H

An Agenda for Future Research

The participation of underrepresented minorities in STEM is multi-dimensional. There is a growing body of research on the social, cultural, psychological, economic, and educational dimensions of broadening participation and success. This is evidenced by, among other things, several recent conferences on understanding interventions that encourage minorities to pursue research careers that showcase the latest research on the problem of women and minority participation as well as on the efficacy of specific interventions. [1]

A selection of promising lines of research and other scholarship on the dimensions of underrepresented minority participation in STEM from the recent past includes:

Economics:

- Samuel L Myers and Caroline Sotello Viernes Turner 2004. "The effects of PhD supply on minority faculty representation," *The American Economic Review: Papers and Proceedings* 94(2)(May 2004):296-301.

---

[1] National Research Council, *Understanding Interventions That Encourage Minorities to Pursue Research Careers*, Washington, DC: National Academies Press, 2007. See http://books.nap.edu/catalog.php?record_id=12022 (accessed February 19, 2010). Anthony L. DePass and Daryl Chubin, eds., *Understanding Interventions That Encourage Minorities to Pursue Research Careers: Building a Community of Research and Practice, Summary of a Conference*, Bethesda, MD: American Society for Cell Biology, 2008. See http://www.understandinginterventions.org/wp-content/themes/simpla_widgetized/files/08Understanding_Interventions.pdf (accessed February 19, 2010).

Sociology:

- Sandra Hanson. 2009. *Swimming Against the Tide: Minority Women in Science*. Philadelphia, PA: Temple University Press.
- Watford, T., M. Rivas, R. Burciaga, and D. Solorzano. 2006. Latinas and the doctorate: The "status"of attainment and experiences from the margin. In J. Castellanos, A. Gloria, and M. Kamimura, eds., *The Latina/o Pathway to a PhD: Abriendo Caminos* (112-133). Madison, WI: University of Wisconsin-Madison Press.
- Willie Pearson, Jr. 2005. *Beyond Small Numbers: Voices of African American PhD Chemists*. New York, NY: Elsevier.
- Sylvia Hurtado et al. 1999. *Enacting Diverse Learning Environments: Improving the Climate for Racial/Ethnic Diversity in Higher Education* (J-B ASHE Higher Education Report Series). San Francisco, CA: Jossey-Bass.
- Kenneth L. Maton, Freeman A. Hrabowski, Metin Ozdemir, and Harriette Wimms. 2008. Enhancing representation, retention, and achievement of minority students in higher education: A social transformation theory of change. In M. Shinn and H. Yoshikawa, eds., *Toward Positive Youth Development: Transforming Schools and Community Programs* (115-132). New York, NY: Oxford University Press.
- Thomas J. Espenshade and Alexandria W. Radford. 2009. *No Longer Separate, Not Yet Equal: Race and Class in Elite College Admission and Campus Life*. Princeton, NJ: Princeton University Press.

History of Science:

- Evelynn Hammonds 2009. *The Nature of Difference: Sciences of Race in the United States from Jefferson to Genomics*. Cambridge, MA: The MIT Press.
- Kenneth Manning. 1985. *Black Apollo of Science: The Life of Ernest Everett Just* New York, NY: Oxford University Press.

Science and Mathematics Education:

- Mary Atwater. 1995. African American female faculty at predominantly white research universities: Routes to success and empowerment. *Innovative Higher Education* 19(4):237-240.
- M. Chang, et al. 2008. The contradictory role of institutional status in retaining underrepresented minority students in biomedical and behavioral science majors. *Review of Higher Education* 31(4):433-464.
- B. C. Clewell, B. T. Anderson, and M. E. Thorpe. 1992. *Breaking the Barriers: Helping Female and Minority Students Succeed in Mathematics and Science*. San Francisco, CA: Jossey-Bass.

- S. Hurtado, N. L. Cabrera, M. H. Lin, L. Arellano, and L. Espinosa, 2009. Diversifying science: Underrepresented student experiences in structured research programs. *Research in Higher Education* 50(2):189-214.
- S. Hurtado, M. K. Eagan, N. L. Cabrera, M. H. Lin, J. Park, and M. Lopez. 2008. Training future scientists: Predicting first-year minority student participation in health science research. *Research in Higher Education* 49(2):126-152.
- Hurtado, S., Han, J.C., Saenz, V.B., Espinosa, L., Cabrera, N., and Cerna, O. (2007). "Predicting Transition ad Adjustment to College: Biomedical and Behavioral Science Aspirants' and Minority Students' First Year of College", Research in Higher Education, 48(7): 841-887.
- E. C. Parsons. 2007. Functioning in two disparate worlds. In K. Tobin and W. M. Roth, eds., *The Culture of Science Education: Historical and Biographical Perspectives*. The Netherlands: Sense Publishers.
- M. Summers and Freeman Hrabowski. 2006. Preparing minority scientists and engineers. *Science* 311 (March 31):1870-1871.

This is a list of promising researchers and scholars, yet this group represents a relatively small cadre, and there is, nonetheless, a need to increase the number of trained researchers whose inquiry is focused on underrepresented minority participation in STEM. A central challenge has been a relative dearth of underrepresented minorities, in particular, formally trained in the history, philosophy, and social study of science. Few nonminority scholars have chosen to write about people of color and STEM, so addressing this dearth of qualified minority researchers is critical to advancing research in the relevant fields. One reason for this dearth is that few underrepresented minorities are enrolled in graduate programs in the elite research institutions that offer the social study of science. In the meantime, too much of the extant literature on underrepresented minorities in STEM has been undertaken by individuals without scientific training.

Along with additional researchers, there can be further advancements in research. Priority areas of inquiry that have been identified by the American Association for the Advancement of Science, Willie Pearson, and Cheryl Leggon include:[2]

- How to create a nurturing institutional and departmental culture that facilitates underrepresented minority success in STEM. New research is needed to better understand the factors that facilitate institutions' embracing of diversity beyond numbers and truly capturing the full benefits that diversity offers.

---

[2] AAAS, 2001. *In Pursuit of a Diverse Science, Technology, Engineering, and Mathematics Workforce: Recommended Research Priorities to Enhance Participation by Underrepresented Minorities*. Washington, DC: AAAS.

- Systematic research to identify those characteristics of the environment and the climate of minority serving institutions (HBCUs, HSIs, TCUs) that sustain and nurture underrepresented minority student interest in STEM education and careers and how predominantly white institutions can adapt those characteristics on their campuses.
- How to develop a critical mass of underrepresented minorities in STEM and the effects of such a critical mass (or lack thereof) on recruitment, social integration, and academic outcomes.
- Understanding of the role of mentoring and mentoring models in STEM education at the high school, undergraduate, and graduate levels.
- Better understanding of the dynamics of creating and sustaining social support networks for students and for faculty.
- Understanding the interaction of gender differences within race and ethnicity in STEM education and careers.
- Understanding the interactions among intervention programs. Existing research rarely distinguishes influence from selection.
- Understanding how participation in multiple intervention programs affects student outcomes.
- Assessing the impact on institutions that have participated in targeted intervention programs, understanding changes in institutional culture; changes in the demographics of students, faculty, and staff; and improvements in the participation (in quantitative and qualitative terms) of underrepresented minorities in STEM.
- Reasons for attrition of underrepresented minorities in STEM along the pathway:
    — Why able and high achieving underrepresented minorities do not enter STEM college majors.
    — Why able and high achieving underrepresented minorities who do enter STEM college majors either do not complete college or switch to other majors.
    — Why more high ability underrepresented minorities do not pursue doctoral education in STEM.
    — Why underrepresented minorities who complete a doctorate in STEM pursue careers outside of academia.
- Identification of the contributions and experiences of eminent underrepresented minority scientists and engineers that can be used to inspire a new generation.

An additional area of social and behavioral research that would benefit from funding is the replication of programs, particularly "best practices" in other environments, answering the question "what works for whom and under what conditions?" Replication is a difficult process and not enough is known about how to do this successfully. Indeed, assessing replicability is a new approach to understanding the improvement of STEM education, both in general and for underrepresented minorities. For example, a variant of the Meyerhoff program at the University of Maryland Baltimore County is now being implemented at Louisiana State University, Cornell University, and Morehouse College. An assessment of these and other efforts at replicability will enhance our knowledge of the effectiveness of potential strategies for using best practices in new contexts.

On the K-12 level, there is a need for longitudinal studies to document the long term impact of Head Start, TRIO, and Upward Bound on achievement in mathematics and science, especially for minorities. Research is needed also to establish the conditions under which AP exam scores lower than 3 relate to college success.